Disaster Risk Reduction

Methods, Approaches and Practices

Series Editor

Disaster risk reduction is a process that leads to the safety of communities and nations. After the 2005 World Conference on Disaster Reduction, held in Kobe, Japan, the Hyogo Framework for Action (HFA) was adopted as a framework for risk reduction. The academic research and higher education in disaster risk reduction has made, and continues to make, a gradual shift from pure basic research to applied, implementation-oriented research. More emphasis is being given to multi-stakeholder collaboration and multi-disciplinary research. Emerging university networks in Asia, Europe, Africa, and the Americas have urged process-oriented research in the disaster risk reduction field. With this in mind, this new series will promote the output of action research on disaster risk reduction, which will be useful for a wide range of stakeholders including academicians, professionals, practitioners, and students and researchers in related fields. The series will focus on emerging needs in the risk reduction field, starting from climate change adaptation, urban ecosystem, coastal risk reduction, education for sustainable development, community-based practices, risk communication, and human security, among other areas. Through academic review, this series will encourage young researchers and practitioners to analyze field practices and link them to theory and policies with logic, data, and evidence. In this way, the series will emphasize evidence-based risk reduction methods, approaches, and practices.

Eiko Wataya · Rajib Shaw

Soft Assets in Smart and Resilient City Development

Springer

Eiko Wataya
Keio University
Fujisawa, Kanagawa, Japan

Rajib Shaw
Graduate School of Media and Governance
Keio University
Fujisawa, Kanagawa, Japan

ISSN 2196-4106 ISSN 2196-4114 (electronic)
Disaster Risk Reduction
ISBN 978-981-97-3372-9 ISBN 978-981-97-3373-6 (eBook)
https://doi.org/10.1007/978-981-97-3373-6

This Springer imprint is published by the registered company Springer Nature Singapore Pte Ltd.
The registered company address is: 152 Beach Road, #21-01/04 Gateway East, Singapore 189721, Singapore

Preface

Globally, approximately 60% of the world's population live in urban areas. Infrastructure is essential for urban development, even more so in developing countries which suffer from rapid population growth, urban overcrowding, unpredictable disasters, negative impact of climate change, deficiency of basic services, slums, and poverty. Hard infrastructure is critical to address many of these challenges. But even more important, at the same time, cities also need to develop and efficiently manage soft assets, incorporating them into project planning and implementation. A balanced integration of soft assets with hard infrastructure fosters sustainable development by prioritizing beneficiaries' needs and reinforces eco-friendly, technology-optimized infrastructure services for maximum benefits on the ground. Soft assets embody a silent architecture that can be seamlessly embedded in project planning and implementation, quietly shaping and producing impact. However, when placed within specific city contexts, whether in the workplace, policy or strategy formulation, institutional systems, procedures, projects, or initiatives, the assets are often unintentionally included, overlooked, or undervalued. Consequently, cities have often accorded soft assets a lower priority to nurture, development, and strategically integrate into their development programs.

Although there are differences between soft (intangible) and hard (tangible) assets, they complement each other. This research discusses how soft assets can be optimized and integrated with hard infrastructure in smart, resilient city planning and development. There are different approaches to achieving the appropriate balance. Over the last 20 years, Multilateral Development Banks acknowledged the significance soft or intangible elements such as human resources, governance, institutional capital, knowledge, partnerships, external relations, and communications in creating a conducive development environment. As a county's development progresses, there is an increasing demand for both knowledge, soft elements, and financial resources to bolster development impacts, but for many developing countries constant challenges remain which impede the achievement of these impacts. These challenges are persistent, regardless of the state of a country or city's development.

The global trend in urban development in developing countries is to prioritize building smart and resilient cities. It is centered around constructing smarter infrastructure to improve the quality of life for citizens and reducing risks of disasters to create more resilient cities. The impacts of climate change, unpredictable disasters, and unforeseen external shocks, such as pandemics, on urban areas have significant implications for the well-being of citizens and overall sustainable growth. Addressing these challenges requires unique comprehensive solutions. However, cities more often prioritize investments in tangible, quantitative infrastructure, frequently overlooking the benefits that qualitative, intangible, soft assets can bring to urban development. Achieving sustainable development requires a deliberate transformative approach and consistent implementation to incorporate soft assets that generate diverse value. Furthermore, a crucial development transformation involves strategic reallocation of a portion of investments from hard infrastructure to soft or intangible assets in city, regional, and national contexts. Establishing an appropriate enabling environment conducive for soft assets within a city's project planning and implementation framework is essential for this purpose.

Many academic and practical studies have examined intangible asset management and measurement. However, tackling this issue is often difficult due to its complexity. Understanding where and how soft assets are incorporated becomes crucial for effective management in the next stage of planning or implementation. Establishing an approach that anticipates value creation is advantageous, and identifying broader issues serves as a critical starting point for improved management in subsequent stages.

This research focuses on the identification, consideration, and utilization of soft assets in the development of smart and resilient cities. It explains the necessary soft asset capital and the key elements for creating value in an urban setting. Drawing insights and data from two India smart city case studies, the research examines the nuanced role of soft assets play in project planning and implementation stages. The primary objective is to recognize and utilize the soft assets which create the foundation for smart, resilient city development. Emphasizing the creation of an enabling environment and sustained value creation over time, the research uses tools such as the 'Blackbox' concept, value creation maps, and 'influential nodes' to reveal a unique approach to data collection and analysis. This involves assessing differences in cities' approaches to the main smart and resilient city elements emphasizing that value creation unfolds in a multi-layered and complex manner. The research also provides an analytical approach for identifying potential areas of value creation given the organic nature of the process and where value emerges dynamically. The key goal is to effectively integrate soft assets into urban planning, recognizing their significant role in creating an enabling environment producing positive results and impacts for smart, resilient cities.

Finally, I wish to acknowledge and thank several individuals and institutions for their valuable advice and support at various stages of this research. Special acknowledgement to the city interviewees from public sector, academia, private NGOs, civil society, and private sector in Bhubaneshwar and Nagpur cities who provided invaluable information and data for the survey; the research advisors from Keio University

(Prof. Wanglin Yan, Prof. Hiroto Kobayashi, Prof. Yoshiaki Miyamoto) for their insightful comments and advice on the research underlying this book; the Resilience Innovation Knowledge Academy India (Ms. Ambika Dabral) for dedicated support to the field research; and Mr. Lester Dally, former World Bank Director and Counselor, for advice on broadening the perspectives of the research and reviewing the final manuscript.

This research introduces a unique approach, but not all aspects are covered leaving several challenges and issues to be explored, creating room for methodological improvement and refinement. Nevertheless, the research seeks to deepen the understanding of this often excluded or invisible value added component contributing to urban development—soft assets.

Fujisawa, Japan

Eiko Wataya
Rajib Shaw

About This Book

The views and findings presented in this book are solely based on the author's independent academic research, findings, and conclusions, and do not reflect the official views of the author's current employer.

Contents

Abbreviations

ABD	Area Based Development
ADMA	Maharashtra State Disaster Management Authority
BDA	Bhubaneswar Development Authority
BMC	Bhubaneswar Municipal Corporation
BSC	*Balanced Scorecard*
BSCL	Bhubaneswar Smart City Limited
BUKC	*Bhubaneswar* Urban Knowledge Centre
CO_2	Carbon dioxide
COVID-19	Corona Virus Disease 2019
CPS	Citizen Participation Survey
EOL	Ease of Living Index
ICLEI	Local Government for Sustainability
ICT	Information and Communications Technologies
IoTs	Internet of Things
ISO	International Organization for Standardization
KPIs	Key Performance Indicators
LGU	Local Government Unit or Administration
MDBs	Multilateral Development Banks
MoHUA	Ministry of Housing and Urban Affairs
MPI	Municipality Performance Index
NMC	Nagpur Municipal Corporation
NMRDA	Nagpur Metropolitan Regional Development Authority
NSSDCL	Nagpur Smart and Sustainable Development Corporation Limited
OECD	Organisation for Economic Co-operation and Development
SDGs	Sustainable Development Goals
SMC	Smart City Mission
SPV	Special Purpose Vehicle
UN	The United Nations

UNDRR	United Nations Office for Disaster Risk Reduction
UNEP	UN Environment Programme
UN-Habitat	The United Nations Human Settlements Programme
UNISDR	United Nations Office for Disaster Risk Reduction
WB	The World Bank

List of Figures

List of Tables

Chapter 1
Introduction

1.1 Background

Cities currently represent about 60% of global population and are the primary growth engines for socio-economic development. With rapid globalization in the last decades, most cities, especially those in developing countries, face multiple challenges. Global issues such as climate change, disaster risks, food insecurity and more recently health pandemics have impacted society, economies, and the environment. To address these challenges, cities are leveraging information, knowledge, and technology to build smarter and resilient cities. Although there is no universal definition of a smart, resilient city, essentially it is an urban spatial concentration where efficient cutting edge technology is employed to make urban service delivery more efficient and increase the competitiveness of a community. It includes improving city governance and administration, enhancing economic activities, optimizing limited resource utilization, addressing environmental degradation, and improving the well-being and quality of life of its citizens.

Rapid population growth and urbanization is inevitable and has become the major driver for developing smart cities. Globally, the urban population is about 4.4 billion or 56% of the total global population. By 2045, the world's urban population will increase by 1.5 times to 6 billion and, by 2050, about 7 out of 10 people will live in cities (World Bank, Urban Development). It is thus critical to plan and develop effective and sustainable cities which provide basic services, infrastructure, affordable housing, and a better living environment for their citizens.

The need is particularly urgent in developing countries which have limited resources to deal with accelerated population growth, higher population density, overcrowding, poverty, and slums. In addition, the speed and scale of the deleterious effects of climate change worldwide is unprecedented. With the greater frequency of disasters and unpredicted crises such as the Corona Virus Disease 2019 (COVID-19), cities need to further refine plans to expand paths to build healthy, safe and sustainable urban environments.

E. Wataya and R. Shaw, *Soft Assets in Smart and Resilient City Development*, Disaster Risk Reduction, https://doi.org/10.1007/978-981-97-3373-6_1

With the dramatic increase in urban populations, building smart, resilient cities is a global trend gaining unstoppable momentum. With accelerating climate change, a smart city aims to use its resources more efficiently, generating lower emissions and reducing its carbon footprint and overall environmental impact (European Commission). Cities also need to prepare for multiple challenges. The ICLEI-Local Governments for Sustainability described the situation as follows: 'a resilient city is prepared to absorb and recover from any shock or stress while maintaining its essential functions, structures, and identity as well as adapting and thriving in the face of continual change…'. Smart and resilient city strategies can have two different focuses: reducing environmental load and strengthening adaptation and transformation. The integration of smart and resilient city approaches is relatively new concept. Both approaches are complementary and necessary to address these new challenges (Baba and Tanaka 2015). At the same time, the evolution of an integrated approach toward a 'smart and resilient city' is required to enhance resilience within a broader smart city strategy (Papa et al. 2015). In summary, smart city approaches as solutions to the challenges and problems associated with urban development have become more important than ever before.

Smart Cities with Hard and Soft Assets

Each city is unique and has its own specific characteristics based on its vision and objectives. The design and development adopted to build a smart city will vary depending on available resources, space, environment design, and available technology. There are several approaches to smart city design. One is an ICT technology-oriented approach (including technology embedded in hardware); another is a people/citizen-centered approach; and the third focuses on top-down and bottom-up initiatives or processes (Ahvenniemi et al. 2017). The city adopts a strategy and process to improve relationships between people and government. Cities also differ in their approaches to hard versus soft infrastructure-oriented strategies and the interlocking relationships between each approach (Angelidou 2014).

The technology and citizen-centered approaches can complement each other. The citizen-centered approach plays a pivotal role in how technology is introduced to cities. First, it serves as a restraint, preventing the excessive acceleration of technology integration. Second and more importantly, the citizen-centered approach also creates opportunities for developing smart and resilient solutions (Wataya and Shaw 2019). The outcome is achieved by adapting to the needs of its citizens and effectively integrating soft assets into the planning, development, implementation, and evaluation processes at multiple levels.

On the first point, to minimize critical risk, citizens cannot be left out of the development process by an excessively hard-oriented IT-led approach. Often technology-centered development ignores the most critical element—effective community integration to ensure citizen welfare and well-being.

On the second point, cities can obtain substantial benefits for smart projects by operationalizing technology and encouraging and utilizing soft assets. With current global challenges, such as climate risks, disasters, pandemics, and other unexpected external shocks, a sustainable smart, resilient city seeks to achieve a more effective use and integration of a range of soft assets to build city resilience in parallel with critical hard infrastructure. Appropriately integrating soft assets in a balanced way with urban infrastructure and building an enabling environment creates resilience, adaptability, and new value for communities and society.

For both approaches, planning and successful implementation of smart city projects requires extensive citizen and stakeholder engagement, technical expertise and the adoption of sound policies, strategies, and institutional arrangements. Hence, the third approach can be considered as a combination of a bottom-up and top-down process. In addition, effective connections, networks, and communications by cities with local communities are key to enable cities to protect citizens and build better and more livable urban environments.

Sing et al. (2023) questioned the aspect of structural inequality in accessing various opportunities such as economic, basic services, and administrative fairness. They argued for the importance of revisiting the choices made toward resilience as they are 'dependent on our capabilities and resource availability and deeply embedded in our social structure and cultural practices' (Sing et al. 2023).

This argument supports the need for soft assets in smart cities to protect the environment and establishing systems to foster resilience. Smart solutions integrating soft assets have become integral to urban development bolstering social, environmental, and economic systems. Technologically integrated hard infrastructure is essential for connectivity and delivery. Increasingly sophisticated tools and applications are being deployed across various stages of smart city development. The choices cities make to enhance 'smartness' also depend on factors such as capacity, the environment, the social system, and urban ecosystem that exists. Using soft assets ensures more effective provision of services and the creation of unique value for cities.

In practical terms, using soft assets is more difficult because they are intangible, indefinable, and often defy qualities of ownership (Haskel and Westlake 2018). Utilization is usually discussed by cities more broadly, but are not always systematically applied at specific local project and community level. Even within the broader context of city planning, the process of value creation of soft assets is not always well understood.

In developing countries, priority is often given to basic physical infrastructure to provide essential and tangible services to citizens. Cities are less likely to consider using soft assets that can strengthen implementation of hard infrastructure due to many factors. They include less rigorous institutional delivery, limited budget, lack of strategic approaches, awareness, knowledge, or understanding of how soft assets create real value for both the city administration and the community. Soft assets have a more granular nature and more often planners find it difficult to incorporate them into both city development overall and within infrastructure projects in particular. Also establishing frameworks to evaluate the impact of incorporating soft assets are challenging.

1.2 Establishing a Value Creation Framework

This research examines how soft assets can be integrated effectively into a smart, resilient city development process, contribute to value creation, and promote livable and sustainable cities. To explore these issues in more detail, two cities from the India Smart City Mission program, launched by the Government of India in 2015, were selected for comparative analysis. The proposal stage of the India Smart City Mission and its implementation stage were compared to identify the different approaches used to incorporate soft assets. Compared to assessing the value of hard assets or physical infrastructure, defining the value of soft assets is more difficult. That is one of the main objectives of this book—to lay out a viable framework to identify and measure the value of these intangible soft assets.

1.2.1 Soft Assets as a Driver for Value Creation

The value-added provided by soft assets has long been recognized. During the development paradigm of the 70s and 80s, the importance of the knowledge-economy was dominant (Miyakawa and Kim 2010). During this period, several variations of the term 'intangibles' developed, i.e. intangible assets, intangible capital, intellectual assets, or even soft assets, and these 'intangibles' were contrasted with hard assets. The term 'intangible' or 'intangible assets' is usually closely associated with the field of accounting, finance, and organizational management. These terms represented an attempt to develop a new model of economic growth incorporating technological development, as well as to encompass new asset classes such as knowledge, collaboration, and information. Intangible assets gained increasing attention in the late 90's when information technologies began to emphasize the importance of economic growth resulting in productivity gains (Miyakawa and Kim 2010). In the business context, this evolution established that long-term expenditure influenced positive economic growth. Also researchers, policy makers, and practitioners concentrated on areas such as accounting, business strategy, management, and investment.

As a result, the concept of intangible assets has become broader and used beyond the business or financial context. More recently, the concept has been applied to the public sector organizational environment (Kaplan and Bower 1999). In this way, the concept used in the public sector can be applied to both urban development and the management context. The benefit of using soft assets not only refers to technological innovation, but also embraces creating value and sustainability through strengthening relations between city authorities and citizens, creating viable communities, outreach, and performance feedback loops.

1.2.2 The Importance of Soft Assets for Smart and Resilient City Development

In developing countries, urban development is largely focused on providing basic infrastructure to improve basic services to citizens. The level and quality of infrastructure implemented depends on the status of development in each country. In general, developed countries implementing smart city policies can employ more innovative approaches based on their existing infrastructure, technologies, and available resources. By contrast, developing countries, with their technological and resource limitations, also have to deal with the challenges of rapid urbanization, overcrowding, and the need to modernize infrastructure.

One of the approaches adopted in Japan is called 'contextualization' or 'personalization' which provides customized services and information to help citizens better prevent and address disasters. Here a smart city authority (or town management company) will take residents' views into consideration, incorporate new services and technologies and continuously support the sustainable evolution of the city or smart town (Fujisawa SST Council 2018). With such personal targeting and feedback systems the city has in place one of the major factors contributing to improving city livability and citizen satisfaction.

Amsterdam has incorporated climate proof infrastructure by prioritizing non-motorized and public transport, permeable green paving, park, and recreational space offering additional benefits for water drainage and improving air quality. The 'Amsterdam Smart City' program is based on collaboration between government, business, and citizens to address urban challenges such as mobility, energy, health care and develop innovative and sustainable solutions creating value. This includes leveraging the knowledge, creativity, and skills of its citizens and local businesses (Home—Amsterdam Smart City). Other cities like Singapore use data effectively in its 'Smart Nation' program where many volunteers have joined to help the community embrace digital technologies bringing their skills and expertise to support the Smart Nation effort (Smart Nation Ambassadors).

These approaches work because effective infrastructure, citizen feedback, and well-being as well as sound policy, project strategies, and environmental enhancements, leverage the advantages of existing technology.

Incorporating these approaches into early city planning is more effective. However, this does not always occur in practice in developing countries as citizen pre-consultations are difficult to facilitate due to lack of human resources, as well as the lower level of citizen awareness and potential involvement in planning and decision-making.

Soft Assets as Invisible Assets

The value and impact created by soft assets through synergetic, spillover, and scaling up effects was always unpredictable and difficult to quantify. This phenomenon creates a sense of 'risk of the benefit of spillover' or unanticipated benefit for decision makers (Haskel and Westlake 2018). The key point about soft assets is to support

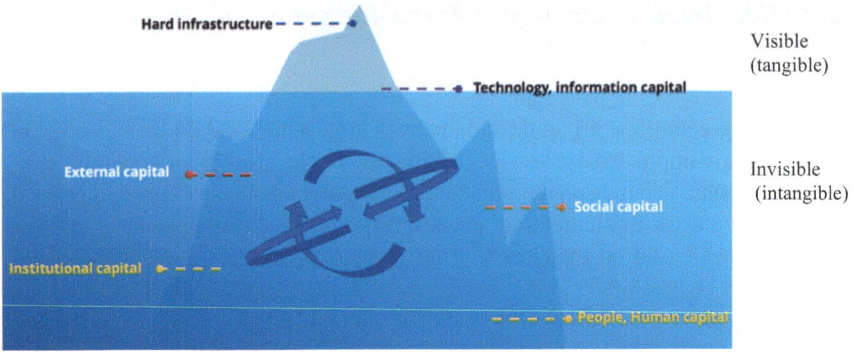

Fig. 1.1 Visible and invisible assets. *Source* Wataya (2023)

hard infrastructure, the visible part, but implementation of soft assets will be less visible as shown in Fig. 1.1. A city can solve the problem by creating an enabling environment to facilitate soft assets and the details are discussed in Chap. 3.

Another example of the benefit of soft assets is provided by multilateral development banks (MDBs) in their assistance to developing countries and underlying clients such as cities. MDBs play a multifaceted role in global development, encompassing policy advice, research and the provision of loans and grants for projects and programs. Their focus extends to addressing the global development agenda, including poverty reduction, climate change, risk management, environmental sustainability, economic sustainability and humanitarian aid, promoting good governance, accountability and developing capacity. Specifically, in terms of soft assistance, MDBs have continued to support strengthening the capacity of developing countries at both individual and institutional levels, enhancing governance through strengthened country systems and networks at regional, city, or community level. Providing solutions, knowledge, and tools to measure performance and impact evaluation have been core areas of support to cities in developing countries. MDBs advice and lending programs provide an application of sound knowledge, solutions, and best practice for city administrations mainstreaming urban development using a solution-oriented, evidence-based decision-making process with innovative, private sector engagement approaches.

The importance of soft assets is supported by the public sector value creation model developed by Kaplan and Bower (1999). Optimal urban development requires increased social, economic, and environmental performance at city level. To achieve this, local authorities need to provide effective, efficient, and quality services to citizens and stakeholders. This is accomplished by city decision makers finding the right balance between hard infrastructure and soft capital development. Cities are also constrained by budgetary factors, the appropriate policy and regulatory mix and the level of knowledge and motivation of decision makers and administrative staff to manage and operate the city efficiently. To reach optimal outcomes, the invisible area shown in Fig. 1.1 represents space to create enabling factors, which produce sound development outcomes. This will be discussed in more detail in Chaps. 3 and 6.

1.2.3 Value Creation Process

Some cities recognize the importance of incorporating soft assets at an early stage, but a major issue for many cities is how to evaluate effective soft asset integration. It is challenging due to the different factors involved. For instance, it is comparable to a black box where clear causal relations and attributions, or linear linkage between inputs and outputs are not immediately obvious (see Fig. 1.2). Also soft asset attributions to outcomes are lower in the early stages, gradually realizing its impact over time (the orange triangle to the left-hand side), but the level of value creation will be higher (the inverted triangle to the right-hand side).

From these characteristics listed in the blue box, the value creation process is very complicated and often difficult to trace in projects and activities at the citywide level. It is important to understand the value creation process (that is, the contents inside the black box) to understand what is required by the city for planning and implementation. This aspect of 'understanding the soft asset black box' is discussed and further explained throughout this book.

Existing research provides an understanding of the process of soft asset value creation which stresses the importance of creating flexible, adaptable spaces that evolve over time to meet citizens changing needs. Jiang and de Rijke (2021) explained

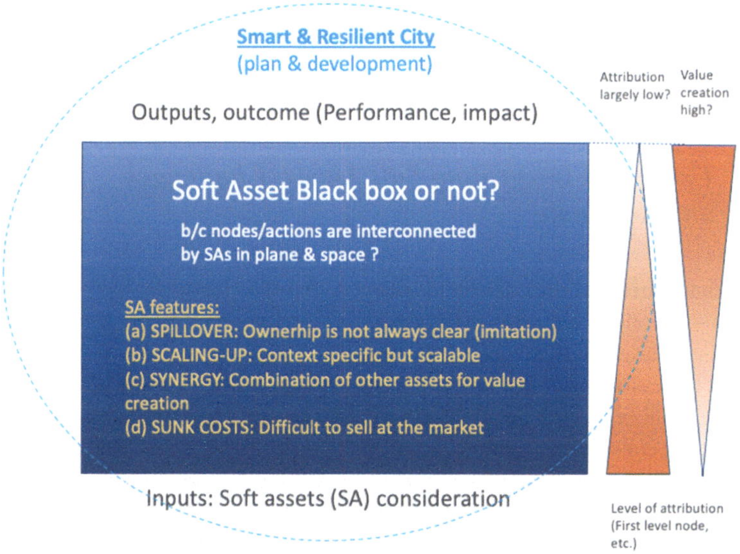

Fig. 1.2 Soft assets in smart and resilient city. *Source* Wataya and Shaw (2022)

the concept called a 'Living structure' that was originally promulgated by Christopher Alexander. They explain how notions, such as 'beauty', are largely judged in a subjective way, but also can be explained objectively (Jiang and Rijke 2021). In their research, they used a computation approach based on a figure of an image rather than the image itself. There are two fundamental governing laws of the living structure concept. First is a scaling law that groups together and emphasizes 'far more smaller things than large ones'. The second is Tobler's law which postulates that everything is related to everything else, but nearer things are more related than distant things (Jing 2019).

Using the concept of living structure from Jiang's study, soft assets can also be assessed not only using an analytical framework, but also through an organic process where value emerges and occurs as shown in Fig. 1.2. Such an organic process is similar to the living structure idea because, in context, soft assets create specific values by synergetic, spill over, and scaling up effects.

Jiang also explained concepts, such as 'wholeness', '... which can be defined mathematically. It exists pervasively in our surroundings; in an ornament, in a room, in a building, in a garden, and in a city'. Alexander previously referred to this phenomenon as the 'quality without a name' (2019). 'Beauty' and the ideas of 'living structure' and 'wholeness' are very similar to how soft assets can be understood, considered and evaluated. Jiang also clearly explained 'Wholeness' in his examination of the living structure concept, '... the evolution process is not simply about adding new centers more correctly, centers are induced by the wholeness. In other words, it is not correct to say: a whole comes from its parts, or a whole consists of parts; it is the wholeness that induces the centers to generate a coherent whole.' (Jiang 2019). This argument supports the assumption that the value of a soft asset should be considered, not only at individual activity and project levels, but also at the more overarching citywide level, especially if they have similar characteristics that Jiang and de Rijke explained as 'living structures'.

1.2.4 Using and Integrating Soft Assets: Challenges and Rationale

Rapid urbanization, climate change, disasters, and other unpredictable risks require effective solutions in smart, resilient city development. With an appropriate balance between soft assets and hard infrastructure investment, cities can provide a higher level of services to their citizens. There is no one-size-fits-all approach, but in general, three points elaborated in the circle in Fig. 1.3 below are key: (i) strategic and systematic soft asset consideration and usage at an early stage; (ii) a realistic and balanced approach to meet each city's development focus; and (iii) institutionalizing mechanisms to manage soft assets for long-term sustainability. This includes integrating a soft asset value creation process into the development process. However, there are also inherent challenges establishing these approaches. This research will examine

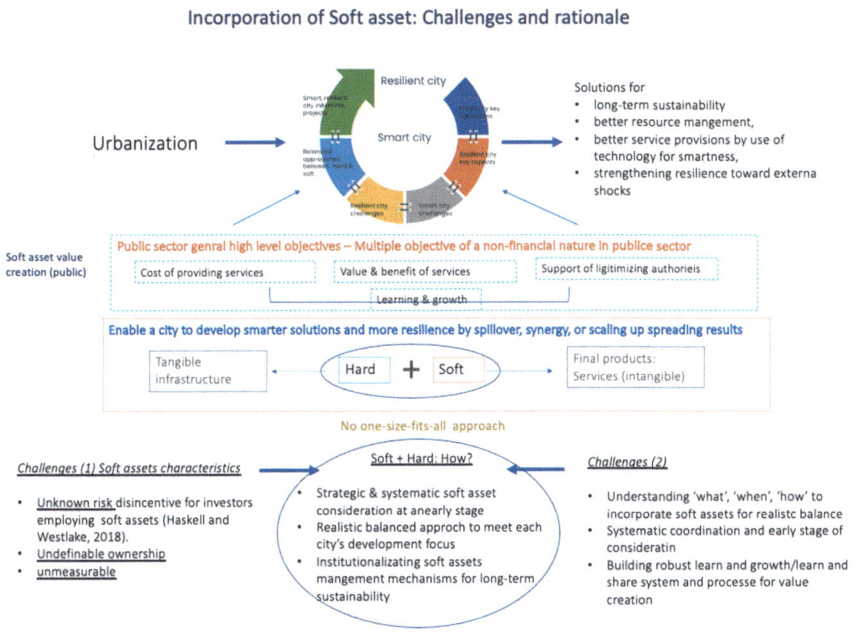

Fig. 1.3 Incorporation of soft assets: challenges and rationale. *Source* Wataya (2023)

these challenges, as well as proposing a value creation system at city level to enhance these approaches.

The next two chapters review in depth four key background topics: (i) urbanization; (ii) integration of smart and resilient city policies; (iii) the state of developing countries' smart city development challenges; and (iv) defining soft assets. Then, to highlight what happens at the city level, the original research uses two case studies from Indian cities which implemented smart city policies mandated at national level. The case of India is a very instructive model to examine. India has implemented one of the most extensive national smart city programs to deal with 'urbanization' by modernizing its urban infrastructure and improving service delivery. Each city proactively incorporated soft assets within this process and lessons can be learned from the experiences of this highly populated country where projects have dealt with 'modernization of infrastructure' and 'urbanization' on a massive scale. Figure 1.3 summarizes some of the challenges and rationale for soft asset incorporation.

1.3 City Case Studies in India

Two Indian cities—Bhubaneswar and Nagpur—were selected as case studies from the India Smart City Mission Program to demonstrate how soft assets were incorporated into their planning and implementation process. As background, the India

Smart City Mission was a major program launched in 2015 by the Ministry of Housing and Urban Affairs of the Government of India to drive economic growth and improve the quality of life of city dwellers combining a two-level approach: area-based needs-focused development, and application of smart solutions and technology at city wide level that leads to Smart Outcomes (Ministry of Housing and Urban Affairs, Government of India).

The lessons learned from these two Indian cities can be easily scaled up and applied to other Indian cities and cities in other developing countries. India, with the world's largest population, is the 5th largest economy in the world with 35% of total population or about 494 million people living in urban areas in 2021 (Figs. 1.4 and 1.5). Basic infrastructure in many cities has failed to keep up with the rapid population growth (Tripathi 2017). The provision of basic infrastructure services is a high priority and serves the principal objectives of the India Smart City Mission program: environmentally friendly smart city development, mitigation of climate risks, disaster prevention and response, and improving city resiliency as a whole.

Each city demonstrates the value and best practice of India's Smart City Mission program: both ranked very high on the Government of India's Ease of Living Index 2020 (Ministry of Housing and Urban Affairs (MoHUA), Government of India) (Bhubaneswar: Rank 2 in 'less than million category'; Nagpur: Rank 25 in 'Million+ category'). Each city has a different vision and focus for smart city development and

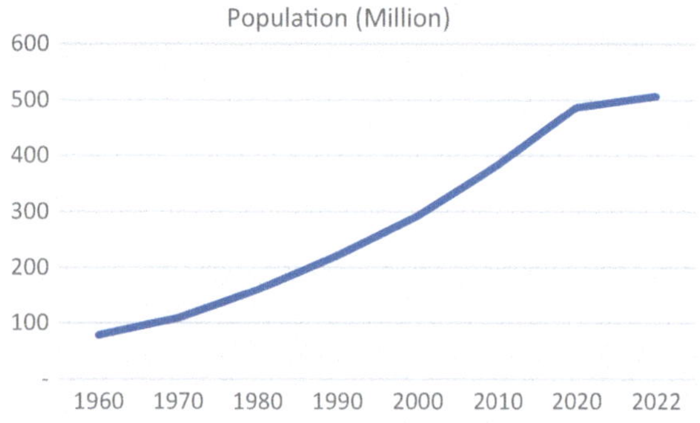

Fig. 1.4 Urban population—India. *Source* WB, https://data.worldbank.org/indicator/SP.URB.TOTL?locations=IN

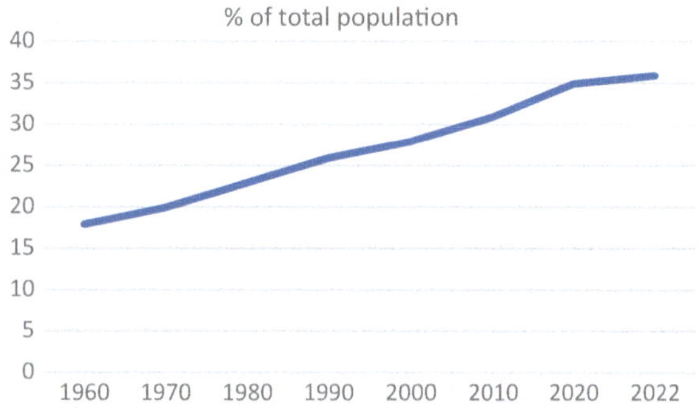

Fig. 1.5 Urban population (% of total population)—India. *Source* WB, https://data.worldbank.org/indicator/SP.URB.TOTL.IN.ZS?locations=IN

unique climate and geological profiles which enables a useful comparison of different resilience policies and results. Also both cities have higher scores than the national average on the Citizen Perception Survey (CPS) score.

According to the EOL, the CPS 'acts as an instrument to validate the citizens' experience in service delivery and assess whether it is consistent with the Ease of Living Index findings. Although the scope of the survey is not exactly the same as the India Smart City Mission (Pan and Area based development), it is a valuable starting point to investigate soft asset integration within cities.

Each city has a positive score on these indicators and are excellent cases to understand and examine: (i) how city planners and stakeholders make decisions to incorporate soft assets during the planning and implementation of the Smart City Mission program; (ii) how a balance between soft and hard assets is achieved; and (iii) how incorporating soft assets creates value, impact and enhances the overall functioning of the Smart City Mission.

1.4 Approaches and Methodology of Analysis

This book was based on original research on 'Integrating Soft Asset in Smart and Resilient Cities in India'. The following questions underlie the research:

- What can be learned from past studies and research about the benefits of soft assets? What approaches or models support the value of using soft assets?
- Using the concepts of soft asset capital and the value creation model, how is utilizing soft assets consistent with achieving smart and resilient city objectives?
- What are the perspectives of city stakeholders about soft asset consideration and how do decision makers and city planners find effective entry points to incorporate soft assets and integrate a value creation process within the smart, resilient city development process?

To address the questions, the research uses both qualitative and quantitative methods to examine, analyze and interpret the information and data gained from a literature review, a field survey, and in-person interviews, as well as visualization of survey results using an open source tool.

References

Ahvenniemi H, Huovila A, Pinto-Seppa I, Airaksinen M (2017) What are the differences between sustainable and smart cities? Int J Urban Policy Plan 60:234–245

Angelidou M (2014) Smart city policies: a spatial approach. Int J Urban Policy Plan 41:S3–S11

Baba K, Tanaka M (2015) A Concept of Resilient city and providing assessment indicators. J City Plan Inst Jpn 50:47–49

Fujisawa SST Council (2018). https://fujisawasst.com/EN/wp_en/wp-content/themes/fujisawa_sst/pdf/FSST-ConceptBook.pdf

Haskel J, Westlake S (2018) Capitalism without capital: the rise of the intangible economy. Princeton University Press, Princeton, New Jersey, USA, pp 221–222

Jiang B (2019) Living structure down to earth and up to heaven: Christopher Alexander. Urban Sci 3:96

Jiang B, de Rijke C (2021) Structural beauty: a structure-based computational approach to quantifying the beauty of an image. J Imaging 7:78

Kaplan RS, Bower M (1999) The balanced scorecard for public-sector organizations, balanced scorecard report—insight, experience & ideas for strategy-focused organization. Massachusetts: Harvard Business School Publishing, Boston, MA, USA

Miyakawa T, Kim YG (2010) Mukeishisan no keisoku to keizaikouka—macro—sangyo kigyo level de no bunseki. RIETI policy discussion paper series 10-P-014, Research Institute of Economy, Trade and Industry (in Japanese). https://www.rieti.go.jp/jp/publications/pdp/10p014.pdf

Papa R, Galderisia A, Majelloa M, Cristina V, Sarettaa E (2015) Smart and resilient cities, a systemic approach for developing cross-sectoral strategies in the face of climate change. TeMA J Land Use Mobil Environ 8:19–49

Sing B, Berger T, Parmar M (edt) (2023) Negotiating resilience with hard and soft city. Routledge, India. https://doi.org/10.4324/9781003299264

The World Bank, Data, Urban population—India. https://data.worldbank.org/indicator/SP.URB.TOTL?locations=IN

The World Bank, Data, Urban population (% of total population)—India. https://data.worldbank.org/indicator/SP.URB.TOTL.IN.ZS?locations=IN

Tripathi S (2017) Relationship between infrastructure and population agglomeration in urban India: an empirical assessment, ADBI Working Paper Series, Asian Development Bank Institute. https://www.adb.org/sites/default/files/publication/301256/adbi-wp731.pdf

Wataya E, Shaw R (2019) Measuring the value and the role of soft assets in smart city development. Cities 94:106–115. https://rajibshaw.org/wpRS/wp-content/uploads/2019/06/Wataya-Shaw-Cities.pdf

Wataya E, Shaw R (2022) Soft assets consideration in smart and resilient city development. Defining and debating on smart cities or smart buildings. Smart Cities 5(1):108–130. https://www.mdpi.com/2624-6511/5/1/7. https://doi.org/10.3390/smartcities5010007

Wataya E (2023) Integrating soft assets in smart and resilient city development in India. Unpublished PhD thesis, Keio University, Japan

World Bank, Urban Development (2024) https://www.worldbank.org/en/topic/urbandevelopment, https://www.worldbank.org/en/topic/urbandevelopment/overview.

Chapter 2
Soft Assets: Review of Recent Literature

2.1 Urbanization, Building Resiliency, and Disaster Prevention

In recent years, urbanization and overcrowding has increased the need for many cities to improve their planning and implementation of programs to enhance social well-being, positive economic outcomes, and environmental sustainability. The world's urban population is now about 4.4 billion or 56% of total global population (World Bank, Urban Development). According to World Bank projections, by 2045, the world's urban population will increase 1.5 times to 6 billion (Figs. 2.1 and 2.2).

Rapid urbanization creates overcrowding and increased poverty which places greater strains on city service provision and has resulted in a vast increase of slums throughout urban areas in developing countries (United Nations (UN) Sustainable Development Goals, Goal 11 Make cities inclusive, safe, resilient and sustainable). The solution lies almost entirely with cities. Cities are potential powerhouses of economic growth and seek to provide opportunities and jobs for its citizens, economic development, and better living standards. Increased migration of people into cities, growing economic development, and industrialization have created larger population densities and congestion (Hayat 2016). This rapid population increase has created challenges for cities which frequently lack resources and consequently suffer environmental deterioration.

The World Health Organization Report (World Health Organization 2021) indicates that from 2015 to 2020, the number of people without properly managed services decreased by 225 million in rural areas, but increased by 32 million in urban areas. Rapid urbanization also adds to critical global challenges such as global warming and climate change. From 1970 to 2010, globally greenhouse gas emissions rose from 30 to 50 tons (Asia–Pacific Economic Corporation 2020).

The increased use of resources by growing population and increased waste generation has resulted in serious ecological consequences in local urban centers extending to regional impacts with resource and waste flowing from cities to the wider area,

E. Wataya and R. Shaw, *Soft Assets in Smart and Resilient City Development*, Disaster Risk Reduction, https://doi.org/10.1007/978-981-97-3373-6_2

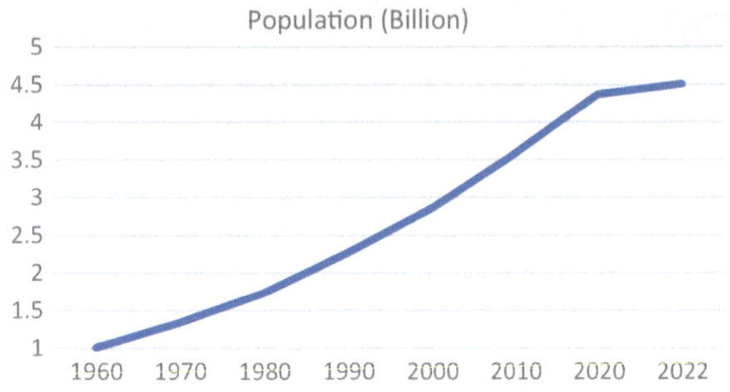

Fig. 2.1 Urban population. *Source* WB Data, https://data.worldbank.org/indicator/SP.URB.TOTL

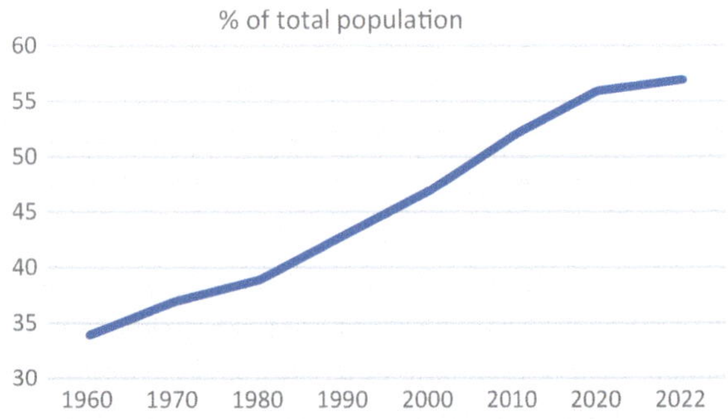

Fig. 2.2 Urban population (% of total population). *Source* WB Data, https://data.worldbank.org/indicator/SP.URB.TOTL.IN.ZS

and exponentially global impacts, including climate change (Satterthwaite 2011). Satterthwaite also argued that, not only has urbanization created these outcomes, but also responsible are higher levels of consumption and more extravagant lifestyles of middle and upper income rural inhabitants.

2.1.1 Reducing risks of crises and disasters and increasing resiliency

Crises and disasters occur at any time and at any level of society. They are natural occurrences and need to be managed by city authorities to mitigate impacts, but also to build a better environment to minimize future crises and risks. On this issue, Michelle (2013) aptly states: '..strengthening the resilience of people and the state to shocks and stress could help protect lives before a crisis hits, reduce potential economic losses, and empower people to take better decisions about the risks..'.

The concept of urban resilience can be examined in different development stages. Kong et al. (2022) enumerated three stages: first, engineering resilience requiring a sound physical system. Secondly, ecological resilience to protect the ecosystem. Thirdly, evolutional resilience which covers long-term development by constantly adjusting city economic, social, and political structures to adapt to frequent disequilibrium. This explains the evolution of a system facing challenges, yet it emphasizes the adaptability within the cycle of innovation and learning. This analysis is widely used for the application and practice of resilience theory (Kong et al. 2022). In this approach, resilience has an organic feature which can be transformed and improved by innovation and learning.

This concept indicates the important rationale for cities to increase resilience by planning and implementing projects to protect against increased risks from extreme weather, disasters, and other external shocks. Although more economic activity and technological innovation is required, a greater use of fossil fuels increases carbon dioxide (CO2) emissions causing global warming and climate change. A sound physical plan is required to reduce CO2 as elevated CO2 levels impact the environment making the city more vulnerable to other climate effects and external shocks. The ability of the city to absorb external shocks is important to address climate change and disaster risks. Disasters have increased more than three times from the 1970 and 80 s (Food and Agriculture Organization of the United Nations 2021). Cities that were poorly planned and developed resulted in inadequate management of rural–urban migration, lower job creation, and poor services reducing, citizen's well-being and making the city more vulnerable. This situation can be reversed by employing smarter solutions to address resilience issues and mitigation risks.

Although resilience is usually associated with events such as extreme weather, disasters or disruptive rural–urban migration, in recent years, the term has expanded to include other external shocks. The impact of multiple crises like the COVID-19 pandemic, climate change, disasters, conflicts, including the Russian invasion of Ukraine and crises in the Middle East, have placed severe strains on many cities. In developing countries, COVID-19 had an major impact on the poorest communities. Hunger and death will continue to increase significantly in urban areas if steps are not taken to reduce poverty and hunger (UN Sustainable Development Goals). In parallel, the associated economic effects have led to higher inflation, economic contraction, and recessions (WBG 2021, 2022).

Different types of natural or man-made disasters or crises have also caused significant harm to people's lives, including loss of assets, property, and socio-economic disruption. Seeking new approaches to resilience is key to address unexpected events beyond municipal control. A recent OECD paper highlighted: 'National, regional, and local governments find they cannot count on following a straight or line as a course of policy action to manage, exit and recovery from the crisis'. One of the key recommendations in the paper emphasizes the importance of regional policy to build more resilient communities and better address future shocks by re-evaluating policy objectives, including urban and rural equilibrium, the climate imperative, the digital divide, the balance between tangible and intangible assets (infrastructure such as broadband, public transport, and social housing, R&D, innovation), citizens' well-being, and the importance of supporting culture (Organisation for Economic Co-operation and Development (OECD) 2021).

2.2 Different Approaches to Smart, Resilient Cities Development

The sustainable city growth (environmental, social, and economic) model provides solutions to address urbanization challenges. However, the model has certain limitations. It aims to achieve socio-economic growth and well-being by using natural resources and services for development in a sustainable manner, enabling future generations to be able to meet their resource needs. Also, to manage a greater level of inputs (i.e. people, services, resources, data, etc.) with fewer resources in a smarter way, using information and communications technologies (ICT) linked to hard infrastructure to strengthen city development. However there often remains a significant digital divide in many cities (Shin, et al. 2021). To address the various urban challenges while promoting sustainable city growth, the smart, resilient city needs a more comprehensive strategy as well as a more holistic approach.

2.2.1 Smart City: Definition and Approaches

The concept of an optimal 'smart city' has no commonly accepted definition in research or practice. The World Development Report (WB 2016) definition is 'a city that leverages the latest in technology and connectivity to make better decisions and achieve the urban aspirations of its residents'. The European Commission definition: 'A smart city uses its resources more efficiently, with less emissions and carbon footprint, minimizing environment impact' (European Commission). Perhaps one of the more comprehensive definitions is from the International Standard Organization's (ISO) that a smart city '.. provides social, economic, and environmental sustainability outcomes. Smart cities respond to challenges such as climate change, rapid population growth, and political and economic instability by fundamentally improving how they engage society, apply collaborative leadership methods, work across disciplines and city systems, and use data information and modern technology to deliver better series and quality of life in the city (residents, businesses, visitors), now and for the foreseeable future, without unfair disadvantage of others or degradation of the natural environment' (ISO/DIS 37122 https://www.iso.org/obp/ui/#iso:std:iso:37122:dis:ed-1:v1:en).

In Japan, the Prime Minister's office explains the characteristics of a smart city as 'solving global issues such as the Sustainable Development Goals (SDGs), through Japan's society 5.0' (Prime Minister's Office of Japan). Smart cities are also providing services tailored to individual citizens using new technologies and various types of data from the public and private sectors as well as upgrading management in various fields, which creates values and sustainable cities and regions that extend beyond 'urban' initiatives to include regional development.

Japan's smart city framework involves three pillars linked to its urban infrastructure development, namely: (i) eco-cities (environmentally symbiotic cities); (ii)

Transit-Oriented Development or TOD; and (iii) Building Disaster-Resilient Cities (Resilient Cities). These pillars underline the strategy for data-driven cities. In parallel, Japan focuses on a human-centered society ('Society 5.0') using digital tools to solve socio-economic issues to improve urban infrastructure. The strategy reflects the central government objective to combine Japan's competency in quality infrastructure and advanced digitization to build smart cities which are more inclusive and conducive to business creating growth with openness and transparency (Prime Minister's Office of Japan).

The core concept of a smart city remains highly heterogeneous, and most studies conclude that cutting-edge technological applications are the most important components in smart city development to create urban efficiencies. Use of technology and real-time data provides a smart city with innovative solutions for the many socio-economic and environmental challenges it faces (Tan and Taeihagh 2020).

2.2.1.1 Diverse approaches

There is no universal definition of smart cities and there are also different approaches to design and implementation. The empirical studies on the topic vary from the examination of the policy level to implementation and challenges, the design of a smart city framework and indicators to evaluate performance of high-end technology, the Internet of Things (IOTs), big data, artificial intelligence and risks in technology integration in smart city development (Neirotti et al. 2014). Given the significant heterogeneity of smart cities, their particular characteristics shape each model of smart city development. Nevertheless, as raised in Chap. 1, three broad approaches to the evolution of the smart city concept are examined.

One is the ICT and technology-oriented hardware approach using technology as a primary driver to improve urban infrastructure and services. The second is a people-oriented (citizen-centered) approach that emphasis that emphasizes the needs and preferences of citizens and communities using a participatory approach. Smart technology-linked hard infrastructure (referred to in this book as 'hard infrastructure[1]) is implemented by the government or service providers through community mobilization and applied to enhance the daily lives of citizens. Here the third approach focuses on top-down and bottom-up initiatives, which aim to improve relationships and decision-making processes between people and government (Neirotti et al. 2014, Angelidou 2014). It also differentiates between hard versus soft infrastructure-oriented strategies and the interlocking relationships between each.

ICT-integrated hard infrastructure is not sufficient alone and is ineffective without soft assets to deliver outputs and ultimately an appropriate level of services to city dwellers. Moreover, hard assets alone do not define a city; it is rather the participation of people and other enabling factors that are vital. Some major different approaches are briefly explained below.

[1] 'Hard infrastructure' means ICT-integrated physical infrastructure since today, most of physical infrastructure are linked to technology.

Hard infrastructure, technology-centric: A smart city needs to distribute limited resources in an effective way and the use of innovative technology for smart infrastructure is indispensable. Smart solutions contribute to making people's life more convenient. In smart city development, technology or ICT has been routinely applied, involving substantial investments in technology-centric projects, but the question is the extent to which local authorities are willing to promote citizen inclusiveness. An exclusive focus on hard infrastructure and technology risks ignores the most critical element—community integration and promoting the welfare and well-being of citizens. The following are some examples of technology-centric hard infrastructure designed for smart city development.

In the smart transport sector in Japan, for example, collecting and translating large amounts of data are utilized to improve city traffic operations. These include mobile-based ITC solutions, including provision of real-time data for public transport services to strengthen public security. In smart solid waste management, the IoTs enable tracking waste in South Korea using radio frequency identification to locate and track containers, identify waste, and verify service delivery. In smart mini-grid renewable energy in Sri Lanka, village hydro projects help the provision of electricity and socio-economic improvements in rural areas through PV (photovoltaics), hydro, wind, and biomass renewable energy technologies. In Mexican smart water management, smart sensors detect available drinking water used for efficient water management. There are several examples in other countries in sectors where smart technology provides significant benefits (Wataya and Shaw 2019).

Technology or digital transformation is very important for smart, resilient city development, but the approach needs to engage all residents and urban stakeholders with city authorities (The United Nations Human Settlements Programme (UN-Habitat) 2020). First, this is because hard infrastructure comprises 'tangible or physical assets' where investment returns are easily measurable. However, hard infrastructure alone cannot deliver optimal 'people-centered' services without human/institutional and other non-physical assets. Second, new quick fix technologies are an easy investment choice for city authorities. However, hard infrastructure is not always appropriate for many cities due to different development objectives, strategies, and resource constraints, including the existence of capacity gaps for planning, managing, operating, and improving city performance. Third, digital technologies have significant potential for sustainable urban development. Therefore, smart cities can provide a solid basis to improve people's lives and build city systems that effectively serve their communities (UN-Habitat 2020).

Citizen-focused, people-centered development: The importance of human and social capital needs to be emphasized, since smart city development relies on a variety of resources, along with innovative and creative approaches, to assist a city build robust social systems to benefit its citizens.

There are some examples where cities can improve the quality of citizens' lives without depending on high-end technologies or popular new hard infrastructure. These use existing infrastructure, improve monitoring, and assist citizens to engage in the decision-making processes of local authorities.

One example is Paris, where old buildings have been converted into multi-purpose use such as providing office space for social enterprises, sharing workplaces, or using public space for both residents and tourists. These facilities aim to make the city more livable and provide a better environment for citizens. Hard infrastructure and technologies are embedded in the social fabric and used to improve human interactions, bring citizens together, solving problems, and reducing barriers to access various city resources.

Another example from the Japan is the Kakogawa smart city, Hyogo prefecture, which illustrates a citizen inclusive approach in its smart city planning stage. The city used a communication technology platform as a central tool to tackle safety and security issues to make the city a 'town chosen by the children-rearing generation' based on the Kakogawa City Town, People, and Work Creation Comprehensive Strategy. Kakogawa launched the project called 'Kakogawa City Project' aiming to solve a regional issue to improve citizen satisfaction and quality of life. Kakogawa has built a stakeholder focused system using a data platform to collect feedback from various areas for analysis. This project has endeavored to establish the best way to create a smart city by proactively engaging citizens. The open data collected is used for the administrative information dashboard and the application called the 'Kakogawa App' for smartphones (Cabinet Office et al. 2021, Fusa-Net 2020).

There are also more cases showing how people are encouraged to engage in different levels of smart city development and also how cities create ways for the public to contribute to innovative solutions to improve the quality of life and services. Although Nigeria faces major challenges digitizing infrastructure, its 'Eyes & Ears project' is designed to combat overspending and under-delivered government infrastructure projects by enabling citizens to monitor and evaluate results with various IT devices and applications. Madrid launched the open-source platform software for the public to engage in the policy-making and spending processes associated with budget proposals supporting new legislation. The platform allowed real-time citizen consultation on council proceedings, debate, and deliberation. Amsterdam started the Chief Technology Office in 2014 using innovative short-term pilot projects in collaboration with local IT businesses and startups, maintaining an open data portal with datasets for each project. In partnering with local academia, the Chief Technology Office plays an important role at the intersection of policy, technology, and design to improve quality of life for citizens (UN-Habitat, 2020).

From these cases, important lessons can be learned, including that technology empowers citizens by facilitating valuable feedback into the decision-making process. They employ distinct approaches to create resilient social systems to enhance citizen's well-being and prioritizing less ambitious ICT infrastructure investments, which can sometimes fail to deliver expected outcomes. Quality of life is improved by a holistic approach and use of soft assets across the economy, environment, and social sectors (Wataya and Shaw 2019). The people, human-centric approach can blend hard and technological solutions to achieve a city's objectives for smart, budget conscious, integrated, cost-effective results, and achieving resource efficiencies. In this way, cities can be environmentally friendly and, sustainable, but also create a positive impact on citizens' well-being and financial situation.

2.2.2 *Resilient City*

'Resilience' covers different urban domains, such as social and ecological resilience, the socio-ecological system, urban hazard mitigation, and community resilience. Resilience is becoming central in re-framing policy strategies with comprehensive cross boundary approaches to address challenging development agendas (Galderisi et al. 2020). Resilience has also a more organic nature which can absorb inputs from innovation and learning. The OECD describes it as follows: 'resilient cities are cities that have the ability to absorb, recover and prepare for future shocks (economic, environmental, social and institutional) and also promote sustainable development, well-being, and inclusive growth' (Resilient Cities, OECD).

The concept of a resilient city is still evolving in the urban context. ICLEI-Local Governments for Sustainability described it as follows: 'a resilient city is prepared to absorb and recover from any shock or stress while maintaining its essential functions, structures, and identity as well as adapting and thriving in the face of continual change'. Building resilience requires identifying and assessing hazard risks, reducing vulnerability and exposure and lastly, increasing resistance, adaptive capacity, and emergency preparedness (ICLEI 2019).

With more specific focus on disaster risk, the United Nations Office for Disaster Risk Reduction (UNISDR) characterizes a resilient city as a city capable of withstanding the impact of hazards through resistance or adaption, which enables a city to maintain certain basic functions and structures during a crisis, and recover from a disastrous event (UNISDR 2012, Making cities Resilient).

The importance of resilient cities reflects some major initiatives by leading institutions such as Making Cities Resilient 2030 by UNISDR which lays out plans for cities to become inclusive, safe, resilient, and sustainable by 2030 (UNISDR, Making Cities Resilient 2030). Also Resilient Cities by CGIAR (Consultative Group for International Agricultural Research) provides research support to strengthen informal urban and the peri-urban agri-food sector, to help improve sustainability, equity and opportunity growth, and to mitigate risks to human and environmental health. The Resilient Cities Congress by ICLEI has provided an international platform to share the latest knowledge, best practices, challenges, and innovations for creating more resilient cities (ICLEI 2019). The 100 Resilient Cities initiatives by the Rockefeller Foundation is designed to help cities build resilience, providing guidance in finance and logistics, resilience strategy development and solutions from experts through a network of private and public organizations and NGOs with access to peer learning from network members (The Rockefeller foundation, 100 Resilient Cities). The definition of resilience used by the Foundation is 'the capacity of individuals, communities, institutions, businesses, and systems within a city to survive, adapt, and grow not matter what kinds of chronic stresses and acute shocks they experience' (The Rockefeller Foundation 2015). This approach also aligns with the transformative definitions of resilience previously discussed.

Past experiences with disasters and unexpected shocks made cities more aware of resilience issues. It is also undeniable that the COVID-19 pandemic also placed cities on heightened alert to provide guides, tools, and solutions for scaling up responses. It also highlighted the robust relationship between the environment and livelihoods (ICLEI 2021, b). There is a demonstrated need for coordinated and systemic multi-hazard prevention and reduction mechanisms within and across all developmental sectors due to increasing global risks and population (Food and Agriculture Organization of the United Nations (FAO 2021). It is also argued that nature-based solutions are needed to deliver solutions to urban populations by providing green places and infrastructure to build resilience and improve adaptation and mitigation measures (The United Nations Environment Programme (UNEP) 2021). The solutions need to be adapted and implemented by a well-trained city administration not hampered by bureaucratic barriers. Also it underscores the importance of comprehensive or holistic approaches toward sustainable and resilient cities.

2.2.3 Integration of Smart and Resilient City Concepts

There are several studies identifying the differences between the two concepts of smartness and resilience. For example, one combines the components and character of a smart city and a resilient city (Arafah, et al. 2018). The other seeks to achieve better integration of the widely pursued strategies. However, successful integration has to be based on a robust scientific approach to provide methodologies and tools to promote cross-sectoral and multi-objective strategies (Papa et al. 2015).

A smart city using ICT provides citizens more flexibility in time and space, mobility, and both physical and virtual access. Such features augment connectivity and enhance a sense of belonging and social cohesion where communities are fully engaged in the process of smart city development. A resilient city improves the quality of life of its citizens with a capacity to absorb shocks and cope with unexpected threats to protect people's lives, assets, and property, maintaining a safe environment, preserving nature, and maintaining a functioning socio-economic system. Preserving citizen safety and security is optimized by the city's indigenous knowledge, development experiences, social ties, robust legal systems, and effective technologies.

The approaches to building smart and resilient cities overlap. In short, smart and resilient city approaches are mutually complementary models with two different emphasizes: reducing environmental load (in the case of a smart city) through the integration of advanced technology and data-driven solution, and strengthening adaptation and transformation addressing challenges and managing disruptions (for a resilient city) (Baba and Tanaka 2015).

Figure 2.3 shows a simplified picture of the integration of two different cities and their common features. Figure 2.4 illustrates the process of smart city development as resilience capacity is improved and absorbed, then readjusted, leading to an evolution of dynamic disequilibrium as part of the smart city development process.

Figure 2.3 shows the overlapping dynamics of smart and resilient cities. Figure 2.4 suggests these items do not occur in precise or predictable order. The key point is

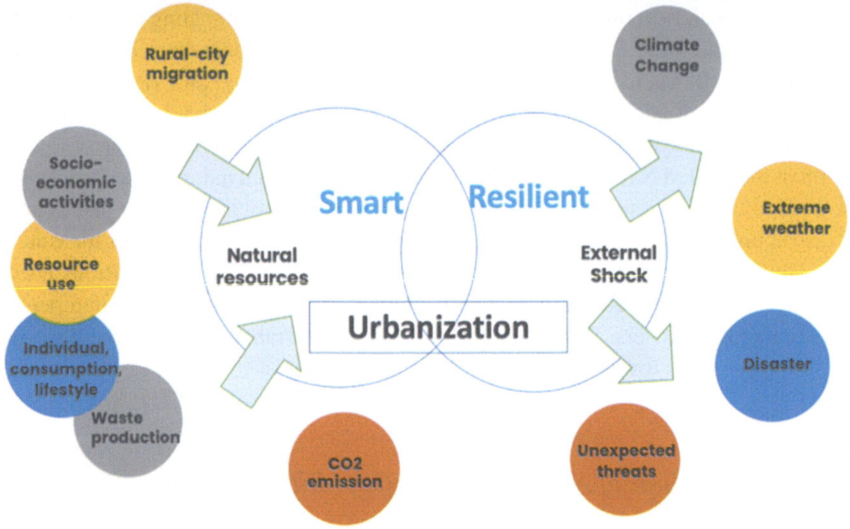

Fig. 2.3 Smart and resilient cities—differences, synergy, and integral parts. *Source* Wataya 2023

Fig. 2.4 Smart and resilient cities—differences, synergy, integral parts, and evolution. *Source* Wataya 2023

that resilience is strengthened through an organic and evolving process, changing, transforming, and absorbing better inputs from innovation, experiences, information, knowledge, and learning. In this regard, full integration of the smart and resilient city is more appropriate than addressing resilience alone. There is more scope to integrate soft assets as enabling factors to build cities, not only by focusing on traditional 'smartness' aspects, but also factoring in 'resilience' as an integral part of comprehensive city development (Papa et al. 2015).

2.3 Smart City and Developing Countries

From the previous literature review and analysis, the discussion of smart and resilient cities involving ICT, digital technology improving citizens' quality of life is largely similar across most countries. However, developing countries have different challenges and priorities compared to developed countries. Figure 2.5 shows the trends and speed of urbanization in selected regions. For the comparison of the speed of urbanization between developed and developing countries, trends in urbanization for less developed regions start in each region with the percentage urbanized in 1950. The percentage of population living in urban areas from 1950 to 2022 are plotted in Fig. 2.5 and Table 2.1 shows the percentage of population increase due to rapid urbanization.

Figure 2.5 and Table 2.1 show the rapid urbanization in the regions of sub-Saharan Africa, Eastern Asia, Western Asia and Latin America and the Caribbean. The most rapid urbanization occurred in Sub-Saharan Africa: the percentage increase is from 11.1 to 42.5%, Eastern Asia with 17.9–66.7% and South-Eastern Asia with 15.6–47.2% over 72 years. From 2000 to 2022, East Asia experienced the most rapid urbanization with a 58.6% increase compared to other regions.

From 2008 to 2019, 153 cities around the world have published official smart city strategies (Berger 2019). Figure 2.6 shows the regions in 2019 that adopted smart city strategies with Europe the most. In Asia, the People's Republic of China and India have the most published strategies (Asian Development Bank 2020). According to Berger, despite 500 cities globally announcing plans for smart city strategies, only

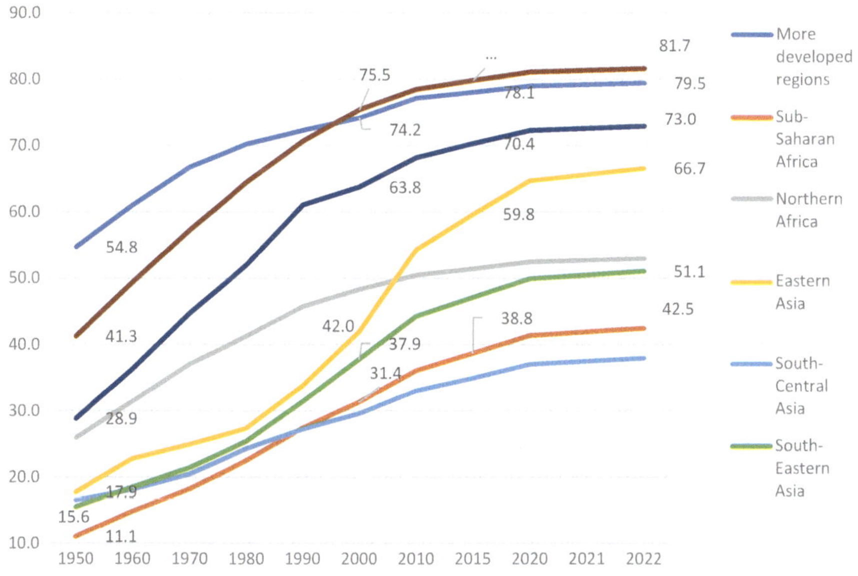

Fig. 2.5 Annual percentage of population: Level and trends of urbanization in selected regions

Table 2.1 Annual percentage of population increase rate of urbanization in selected regions

						(%)
	1950	2000	2015	2022	% increase rate (1950–2022)	% increase rate (2000–2022)
More developed regions	54.8	74.2	78.1	79.5	45.1	7.1
Sub-Saharan Africa	11.1	31.4	38.8	42.5	283.1	35.4
East Asia	17.9	42.0	59.8	66.7	273.0	58.6
Western Asia	28.9	63.8	70.4	73.0	152.6	14.3
Latin America and the Caribbean	41.3	75.5	79.9	81.7	97.8	8.2
South Eastern Asia	15.6	37.9	47.2	51.1	228.1	34.9

Source Annual Percentage of Population at Mid-Year Residing in Urban Areas by region, subregion and country, 1950–2050, https://population.un.org/wup/Download/, United Nations Department of Economic and Social Affaires Population Division), United Nations, Department of Economic and Social Affairs, Population Division (2018). *World Urbanization Prospects: The 2018 Revision,* https://esa.un.org/unpd/wup/Publications

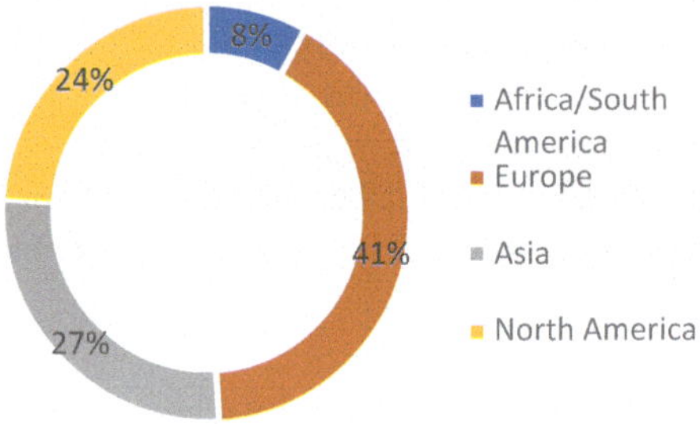

Fig. 2.6 Percentage of official smart city strategy publication. *Source* Berger, 2019

49 strategies were actually published. The low number of published strategies can be attributed to a lack of capacity and resources or according to a lower priority in countries where establishing basic infrastructure and services was more urgent (Asian Development Bank 2020).

2.3.1 Smart city development in developing countries

To fully benefit from smart, resilient city development, many developing countries face enormous challenges to implement the required human, legal, and regulatory reforms. Also governments need to deliver basic infrastructure to meet the needs of citizens, raise revenues to provide better services, enact regulatory frameworks to mitigate the technological risks involved, develop human and institutional capital, ensure digital inclusiveness and promote environmental sustainability (Tan and Taeihagh 2020). Contrast the situation of developed countries where basic infrastructure is already deployed with the necessary systems and institutions established to enable smart city development.

Tan and Taeihagh (2020) conducted a literature review which surveyed key factors that motivate developing countries to develop and support smart cities. In summary these were (i) improving government efficiency in public service delivery; (ii) improving citizen quality of life; (iii) promoting inclusive governance; and (iv) inclusiveness of vulnerable and disadvantaged populations. Developing countries also have different priorities in their smart city development. For smart cities, technological infrastructure is a priority. Other major areas of focus are financing capacity, regulatory development, including promulgating a clear set of laws, regulations, and policies that guide smart city development; and human capital development, including technical capacity and motivation to implement IT-enabled services. Lastly, citizen engagement is important that embraces citizen empowerment, building ownership, and encouraging behavioral changes to create strong social capital.

On the other hand, there are also implementation challenges. Besides budget constraints and lack of investment in basic infrastructure, there are many areas where soft assets should be employed to fill the gaps, including: (a) technology-related infrastructure readiness; (b) governance frameworks and regulatory safeguards for smart cities; (c) skilled human capital; (d) inclusivity; and (e) citizen participation. In addition, there are problems such as fragmented governance, environmental issues, lack of technological knowledge, and capacity of citizens.

These challenges are not unique to developing countries as many countries also lack soft asset capital and capacity. Tan and Taeihagh (2020) argued that cities face different challenges due to availability of foreign investment resources, an adequate regulatory environment, and appropriate political systems. It is difficult to generalize, but important to carefully study each unique city context to gauge the strategic challenges required for optimal smart city development.

2.4 Key Components for a Smart City and Some Assessment Frameworks

Despite varied definitions, the different approaches and performance assessments of smart cities in different development stages need to be reviewed. Cities are growth centers and melting pots for social change and significant challenges exist, including environmental degradation, poverty, inequality, and social violence (University of Birmingham 2016). How can the key elements of a smart, resilient city be identified and assessed? Some assessment frameworks are briefly examined below.

2.4.1 Characteristics of a smart city

Giffinger et al. (2007) identified six characteristics of smart cities which provide a framework for indicators to assess smart city performance: (i) 'Smart Economy' that includes all economic activities and connections around competitiveness; (ii) 'Smart people' that includes not only the level of education or qualification but also interaction and connectivity with externalities; (iii) 'Smart Governance' that includes political participation, service for citizen and administrative functionality; (iv) 'Smart Mobility' that includes access to local and international information on smart transport network; (v) 'Smart Environment' which refers to green space, low pollution, and resource management to prevent environmental degradation; and (vi) 'Smart Living' that includes quality of life, health, safety, culture, affordable housing, and others factors supporting citizens' well-being.

These characteristics represent both sides of the smart city spectrum—easily identifiable hard physical infrastructure and soft elements contributing to effective implementation such as people-oriented governance and communications, all essential building blocks to establishing smart cities. These characteristics are also covered later in Chap. 3 in the discussion of soft assets value elements.

2.4.2 Various assessment frameworks

Assessing the outcomes of integrating soft assets is not always the same as measuring smart city performance based on project sustainability by project-specific-based indicators or more extensively based on the indicators like those in ISO 37120. Measurement comprises nineteen economic themes from basic needs such as water, wastewater, sports, culture, and others under three different pillars consisting of core, supporting, and profile indicators (International Organization for Standardization (ISO) 37,120). The set of indicators is very general. Therefore when Berman and Orttung (2020) applied them to their study of Arctic cities, they found that the ISO 37120 standards lack more regional specific contexts. Since the ISO covers a broader scope of city implementation, assessing outcomes of soft asset integration can provide a complementary approach.

The United Nations Sustainable Development Goals (UN SDGs) 11 refers to making cities and human settlements inclusive, safe, residential, green, and sustainable that focus on citizens' well-being and their quality of life. Some city-level indicators focus on sectors or projects, systems or processes, but the majority are still conventional indicators designed to measure improvement by calculating project-based outputs reflecting performance outcomes.

Other indicators such as the EcoCity Framework comprise 18 standards in four categories, namely, 'urban design', 'big geo physical conditions', 'social cultural conditions', and 'ecological imperatives' (The EcoCity Standards). The indicators were developed around how cities can make better progress toward creating conditions for ecocity status.

On climate change, the Government of India launched the Climate Smart Cities Assessment Framework 2.0 (Ministry of Housing and Urban Affairs, Government of India). The framework provides for 100 Indian smart cities to implement holistic, climate responsive development, to raise awareness and foster behavioral or mind-set changes while engaging in smart city development initiatives or projects.

As Table 2.2 shows the Climate Smart Cities Assessment Framework 2.0 covers 5 key pillars associated with climate change and environmental degradation. They are (i) urban planning, green cover, and biodiversity; (ii) energy and green building; (iii) mobility and air quality; (iv) water management; and (v) waste management. As the government requires preparation of the city assessment, this tool can be used to conduct self-assessment. Cities conducting the assessment observed that there are peer-learning opportunities through knowledge sharing platforms which motivated them to collectively address climate change impacts. The government also plans to create feedback loops to improve the indicators and assessment methodology and strengthen results measurement (Sustainable Urban & Industrial Development).

There are also other smart city indicators and assessment frameworks. Among those briefly reviewed, some have similar or overlapping areas covering overall sustainability, urbanization, habitat focuses, ecocity development, and climate city development. Focus by cities on these areas can become a source of growth and value creation. These are the key core soft asset value elements for smart, resilient cities. The comparison table is presented in Table 2.3.

These indicators provide further guidance on how the process of soft assets incorporation can be addressed. Regardless of the breakdown of sectors or pillars, these are areas where balanced approaches between hard and soft assets are required to achieve smartness and resiliency, as well as sustainability. The smart city characteristics, factors, and assessment areas reviewed in this section form an essential part of the foundation for identifying key soft assets value elements in smart cities.

2.5 How Soft Assets Facilitate Sustainability

Quality infrastructure is critical for sustainable city development. An excellent initiative has been undertaken by the Government of Japan—the Partnership for Quality Infrastructure, Investment for Asia's Future (Ministry of Foreign Affairs, Government of Japan 2015). With accelerating population growth and overcrowding, urban communities have demanded from city authorities improved efficiency from brick and mortar hard infrastructure and basic services. Equally important, citizens expect less visible or intangible benefits from service delivery such as a better quality of life and city livability, which is the final outcome of effective integration of 'intangible or soft' assets.

Table 2.2 Climate smart city assessment framework

Climate Smart City Assessment Framework	Urban Planning, Green Cover & Bio diversity	Rejuvenation & Conservation of Water Bodies & Open Areas
		Proportion of Green Cover
		Urban Biodiversity
		Disaster Resilience
		City Climate Action Plan
	Energy and Green Building	Total electricity consumption in the city
		Total electrical energy in the city derived from renewable sources
		Fossil fuel consumption in the city
		Energy efficient street lighting in the city
		Promotion of green buildings
		Green building adoption
	Mobility & Air Quality	Clean Technologies Shared Vehicles
		Availability of Public Transport
		Percentage of coverage of Non-Motorized Transport network (pedestrian and bicycle) in the city
		Level of Air Pollution
		Clean Air Action Plan (Planning and Implementation)
	Waste Management	Landfill/dumpsite Scientific Remediation
		Scientific Landfill availability & operations
		Extent of Wet Waste Processed
		Construction & Demolition (C&D) waste management
		Extend of dry waste recovered & recycled
		Waste minimization initiatives undertaken by the City

(continued)

Table 2.2 (continued)

	Water Management	Energy-efficient wastewater management system
		Energy-efficient water supply system
		Flood/water stagnation risk management
		Wastewater Recycle and Reuse
		Extent of non-Revenue Water
		Water Resources Management

Source Ministry of Housing and Urban Affairs, Government of India (2021a), Climate Smart Cities Assessment Framework 2.0 Cities Readiness Report https://niua.in/csc/assets/pdf/key-documents/Cities-Readiness-Report.pdf

The way soft assets are managed in city development can result in maximum impact in the long term and become the foundation of quality infrastructure in the city. It ensures value for money and reduces life-cycle costs. As an example, the Government of Japan's 'Quality Infrastructure Investment' program comprises: (i) cost-effectiveness in the long run; (ii) enhancing connectivity to develop local skills and improving people's lives; and (iii) customized support for each country's development plan. Soft assets play a critical role here to improve the quality of infrastructure.

To better understand this, integrating soft assets in a dynamic way transforms them into valuable knowledge assets. For example, where some soft assets are considered knowledge or intellectual assets, they create a pool of knowledge which build intellectual values in an organization. A clear illustration of the value of soft assets can be seen from the extensive experiences of the World Bank and other multilateral development banks integrating value creating systems into their projects and operations. Some cases below show the benefits and challenges in establishing and managing a framework for measuring intangible assets. This aspect is further examined in more in Chap. 3.

More than twenty years ago, the World Bank announced a new vision for a 'Knowledge Bank', using development knowledge and expertise as a driving force to help developing countries build capacity and identify operational solutions. The approach evolved from traditional development banking business—offering a package of concessional lending/investment in hard infrastructure and policy recommendations to developing countries—to knowledge-bank solutions emphasizing knowledge and expertise or soft assets as powerful tools for achieving desired development outcomes. Now the concept of the knowledge-bank using soft assets is widely accepted and implemented by many other multilateral development banks and bilateral aid institutions. One of the challenges is how to properly evaluate this knowledge paradigm. An excellent example is 'The development of the knowledge bank and

Table 2.3 Comparison of key themes, sectors, pillars under different assessment indicators and framework

	ISO37120	UN SDG 11	EcoCity	Government of India
Purpose	Urban sustainability	Making cities and human settlements inclusive, safe, resident, and sustainable, or city levels that focus on the citizens' well-being, their quality of life, green cities, etc	City can make better progress toward ecocity status or conditions	Climate Smart Cities Assessment Framework
Indicators	19 themes with 3 different pillars: core, supporting and profile indicators Indicator	9 Indicators by 2030	5 kay categories with 18 indicators	5 key pillars with 28 indicators
Key theme, sector, pillars	Economics	Target 11.1 By 2030, ensure access for all to **adequate, safe and affordable housing and basic services and upgrade slums**	Urban design	Urban planning, green cover & biodiversity
	Education	Target 11.2: By 2030, provide access to **safe, affordable, accessible, and sustainable transport systems for all**, improving road safety, notably by expanding public transport, with special attention to the needs of those in vulnerable situations, women, children, persons with disabilities and older persons	Big geo physical conditions	Energy and green building;
	Energy	Target 11.3: By 2030, enhance inclusive and sustainable **urbanization and capacity for participatory, integrated and sustainable human settlement** planning and management in all countries	Social cultural conditions	Mobility & air quality
	Environment and climate change	Target 11.4: Strengthen efforts to **protect and safeguard** the world's cultural and natural heritage	Ecological imperatives	Water management

(continued)

Table 2.3 (continued)

	ISO37120	UN SDG 11	EcoCity	Government of India
	Finance	Target 11.5: By 2030, significantly r**educe the number of deaths and the number of people affected and substantially decrease the direct economic losses relative to global gross domestic product caused by disasters**, including water-related disasters, with a focus on protecting the poor and people in vulnerable situations		Waste management
	Governance	Target 11.6: By 2030, reduce the adverse per capita environmental impact of cities, including by paying **special attention to air quality and municipal and other waste management**		
	Health	Target 11.7: By 2030, provide **universal access to safe, inclusive and accessible, green and public spaces**, in particular for women and children, older persons, and persons with disabilities		
	Housing	Target 11.a: Support positive economic, **social and environmental links between urban, peri-urban and rural areas** by strengthening national and regional development planning		
	Population and social conditions	Target 11.c: Support least developed countries, including through financial and technical assistance, in **building sustainable and resilient buildings utilizing local materials**		
	Recreation			

(continued)

Table 2.3 (continued)

	ISO37120	UN SDG 11	EcoCity	Government of India
	Safety			
	Social waste			
	Sport and Culture			
	Telecommunications			
	Transportation			
	Urban/local agriculture and food security			
	Urban Planning			
	Waste water			
	Water			

Source Berman and Orttung 2020; UN SDG indicators, https://unstats.un.org/sdgs/metadata/?Text=&Goal=11&Target_=_; https://ecocitystandards.org/framework/; MoHUA, Climate Smart Cities Assessment Framework, 2.0, https://smartcities.gov.in/climatesmart_cities

the goal of encouraging indigenous knowledge are hampered by pressures to show a tangible result' (Cohen and Laporte 2004). It was a major challenge for the World Bank to justify investment in intangible assets and knowledge products to increase the quality and quantity of development outcomes and which were subsequently followed by other institutions. Sound evaluation frameworks were developed by each organization, such as the United Nations Development Group (United Nations 2017), but 'difficulties' remain. National and city governments have a challenge to show clear evidence that investment in soft assets creates positive outcomes and impact for their beneficiaries.

A study on Eurasian countries highlighted similar challenges. They succeeded in natural resources management, but failed because institutions were too weak to deliver effective public services, education, and other key service areas which had severe impacts on productivity and growth. This is because the Eurasian countries focused on extracting and depleting natural resources rather than improving the efficiency of investment in public goods and prioritizing intangible assets. To strengthen development, countries urgently needed to invest in intangible assets (Gill et al. 2014).

Since 1980, devising methods to measure soft assets has been an important topic so the growth of business activities can be accurately captured. Technology development created a paradigm shift in business investment. A focus on the importance of intangible assets has gained more acceptance due to the changing business environment and increased attention to the benefits of the knowledge-economy. In the public sector context, local authorities invariably give a lower priority to soft assets than hard assets as their direct influence on outputs, outcomes, and performance are more difficult to measure than hard infrastructure investment. However, in the smart city context, there is increasing focus on the importance of monitoring, evaluation, and learning during different stages of the smart city development process. A good example is one of the key processes used by India to create its smart cities (MoHUA, Government of India 2021b).

References

Angelidou M (2014) Smart city policies: A spatial approach. Int. J. Urban Policy Plan. 41:S3–S11

Arafah Y, Winarso H, Suroso DSA (2018) Towards Smart and Resilient City: A Conceptual Model. In: The 4th Plano Cosmo International Conference, IOP Conf. Series: Earth and Environmental Science, vol 158, 012045. IOP Publishingpp 6–16. https://doi.org/10.1088/1755-1315/158/1/012045

Asia-Pacific Economic Corporation (APEC) Telecommunication and Information Working Group, (2020) Recommendations for Implementation of Smart Sustainable City Information and Communication Technology Infrastructures in the APEC Region, Asia-Pacific Economic Corporation Secretariat

Asian Development Bank, Yoon SY, Lee HS, Zelt T, Narloch U, Aguire E (2020) Smart City Pathways for Developing Asia, An Analytical Framework and Guidance, ADB sustainable Development Working Paper Series, No 71, December 2020, 1–9

Baba K, Tanaka M (2015) A Concept of Resilient city and providing assessment indicators. J City Plan Inst Jpn 50:47–49

Berger R (2019) Think: act, Smart city strategy INDEX 2019. The Smart City Breakaway, Roland Berger GMBH, Munich, Germany, pp 3–19

Berman M, Orttung WR (2020) Measuring progress toward urban sustainability: do global measures work for arctic cities? Sustainability 12(9):3708. https://doi.org/10.3390/su12093708

Cabinet Office, Ministry of Internal Affairs and Communications, Ministry of Economy, Trade and Industry, Ministry of Land, Infrastructure, Transport and Tourism, Smart City Public-Private Partnership Platform (2021) Smart City Guidebook (Outline). Cabinet Office, Government of Japan. https://www8.cao.go.jp/cstp/society5_0/smartcity/00_scguide_eng_ol.pdf (Ver. 1.01, in English), https://www8.cao.go.jp/cstp/society5_0/smartcity/00_scguide_s.pdf (Ver.1.0, in Japanese)

Cohen D, Laporte B (2004) The evolution of the knowledge Bank, https://www.semanticscholar.org/paper/The-Evolution-of-the-Knowledge-Bank-Cohen-Laporte/727fd4c55e0d9624b04945eca5eaf1a6bef6e024

Food and Agriculture Organization of the United Nations (2021) The Impact of disasters and crisis on agriculture and food security, 38–39

Fswa-Net, Smart cities have Begun in Japan as well! (2020). https://fswa-net.com/news/smart-cities-have-begun-in-japan-as-well

Galderisi A, Limongi G, Salata KD, (2020) Strengths and weaknesses of the 100 Resilient Cities Initiative in Southern Europe: Rome and Athens' experiences. City Territ Arch 7:16. Springer Open. https://doi.org/10.1186/s40410-020-00123-w

Giffinger R, Fertner C, Kramar H, Kalasek R, Pichler-Milanovic N, Meijers E (2007) Smart cities, ranking of European medium-sized cities. Final report, October 2007, Delft University of Technology. https://www.smart-cities.eu/download/smart_cities_final_report.pdf

Gill IS, et al (2014) Diversified development, making the most of natural resources in Eurasia. The World Bank. pp 319–340

Hayat, P (2016) Smart cities: A global perspective, India quarterly. J Int Aff 72(2), Sage journals, 177–191

Kong L, Mu X, Hu G, Zhang Z (2022) The application of resilience theory in urban development: a literature review. Environ Sci Pollut Res 29:49651–49671

ICLEI (2019) Resilient Cities, thriving cities: The evolution of urban resilience. ICLEI–Local Governments for Sustainability, Bonn Germany, https://e-lib.iclei.org/publications/Resilient-Cities-Thriving-Cities_The-Evolution-of-Urban-Resilience.pdf

ICLEI, (2021) Resilient cities action package 2021 (ReCAP 21). ICLEI–Local Governments for Sustainability, Bonn Germany. https://resilientcitiesnetwork.org/downloadable_resources/Programs/ReCAP21_Guide.pdf

ICLEI South Asia (2021) Climate Resilient City action plan—Nagpur, November. https://southasia.iclei.org/wp-content/uploads/2022/04/2.-Climate-Resilient-City-Action-Plan-Nagpur-Report-Low-Res_compressed.pdf (accessed on 8 October 2023)

Ministry of Foreign Affairs, Government of Japan (2015) Partnership for quality infrastructure. Investment for Asia's Future, https://www.mofa.go.jp/files/000081298.pdf

Ministry of Housing and Urban Affairs, Government of India (2021a), Climate Smart Cities Assessment Framework 2.0 Cities Readiness Report https://niua.in/csc/assets/pdf/key-documents/Cities-Readiness-Report.pdf

Ministry of Housing and Urban Affairs, Government of India (2021b) Making a city smart: learnings from the smart cities mission a workbook for decision makers. https://smartnet.niua.org/sites/default/files/resources/making_a_city_smart_mar2021.pdf

Mitchell A (2013) Risk and Resilience: From Good Idea to Good Practice, A Scoping study for the Experts Group on Risk and Resilience, Organization for Economic Co-operation and Development, Working Paper 12/2013, OECD Publishing, Paris https://doi.org/10.1787/5k3ttg4cxcbp-en

Neirotti P, De Marco A, Cagliano AC, Mangano G, Scorrano F (2014) Current trends in Smart City initiatives: Some stylised facts, Department of Management and Production Engineering, Politecnico di Torino, Corso Duca degli Abruzzi 24, 10129 Torino (TO), Italy. Cities: Int J Urban Policy Plann 38, 25–36

OECD (2021) OECD Policy Responses to Coronavirus (COVID-19). The territorial impact of COVID-19: Managing the crisis and recovery across levels of government. https://doi.org/10.1787/a2c6abaf-en

Papa R, Galderisia A, Majelloa M, Cristina V, Sarettaa E (2015) Smart and Resilient cities, a systemic approach for developing cross-sectoral strategies in the face of climate change. TeMA J Land Use Mobil Environ 8:19–49

Satterthwaite D (2011) How urban societies can adapt to resource, shortage and climate change. Phil Trans R Soc A 2011(369), 1762–1783, https://doi.org/10.1098/rsta.2010.0350

Shin SY, Kim D, Chun SA (2021) Digital Divide in Advanced Smart City Innovations. Sustainability 13:4076. https://doi.org/10.3390/su13074076

Tan YS Taeihagh A (2020) Smart City Governance in Developing Countries: A Systematic Literature Review. Sustainability 12, 899:1–30. https://doi.org/10.3390/su12030899. https://papers.ssrn.com/sol3/papers.cfm?abstract_id=3526607

The Rockefeller Foundation, Arup 2015, City Resilient Index supported by the Rockefeller Foundation. https://www.arup.com/projects/city-resilience-index

The United Nations Office for Disaster Risk Reduction. Making Cities Resilient Report 2012; 2012 UNISDR

The United Nations Development Group (2017) Capacity Development, UNDAF Champion Guidance. https://unsdg.un.org/resources/capacity-development-undaf-companion-guidance

The United Nations, Department of Economic and Social Affairs, Population Division (2018) World Urbanization Prospects: The 2018 Revision. https://esa.un.org/unpd/wup/Publications

The United Nations Human Settlement Programme (UN-HABITAT) (2020) Centering People in Smart Cities, A playbook for local and regional governments. United Nations Human Settlement Programme (UN-Habitat) https://unhabitat.org/sites/default/files/2021/11/centering_people_in_smart_cities.pdf

The United Nations Environment Programme, (2021) A working paper for the G20, Smart, Sustainable and Resilient Cities: the Power of nature-based Solutions

The United Nations Office for Disaster Risk Reduction, Making Cities Resilient 2030 https://mcr2030.undrr.org/who-we-are

University of Birmingham, UK Aid, Governance · Social Development · Humanitarian · Conflict (2016), Urban governance. Topic guide, DIFID, November 2016, https://gsdrc.org/wp-content/uploads/2016/11/UrbanGov_GSDRC.pdf

Wataya E, Shaw R (2019) Measuring the value and the role of soft assets in smart city Development. Cities 94, 106–115. https://rajibshaw.org/wpRS/wp-content/uploads/2019/06/Wataya-Shaw-Cities.pdf

Wataya E (2023) Integrating soft assets in smart and Resilient city development in India. Unpublished PhD thesis, Keio University, Japan

World Bank, Green Resilient, and Inclusive Development (2021) World Bank, Washington, D.C. https://openknowledge.worldbank.org/handle/10986/36322

World Bank Group (2016) Sector focus 4, smart cities, world development report 2016: Digital Dividends, World Bank Report 2016. World Bank Group, Washington, D.C. https://www.worldbank.org/en/publication/wdr2016

World Band Group. Global Economic Prospects (2022) A world bank group flagship report. World Bank, Washington, D.C. https://openknowledge.worldbank.org/bitstream/handle/10986/36519/9781464817601.pdf?sequence=10&isAllowed=y

World Health Organization, the United Nations Children's Fund (2021), July, Progress on Drinking Water, Sanitation and Hygiene, 2000–2020, Five Years into the SDGs. https://www.who.int/publications/i/item/9789240030848

Chapter 3
Defining Soft Assets

3.1 Characteristics of Soft Assets

Soft assets are powerful elements which create value when combined with other hard and soft assets. Some soft assets which provide value can be easily identified, but others are more difficult to quantify or classify. The value proposition of soft assets will often depend on the specific city context, as well as the type of assets utilized to align with a private firm's vision and strategy (Kaplan and Norton 2004a, b). In the business context, soft assets often refer to productive capacity. Furthermore, Kaplan and Norton argued that intangible assets seldom exert a direct impact on financial performance, but operate indirectly through intricate chains of cause-and-effect. However, they lack the capacity to independently generate value (Kaplan and Norton 2004a, b).

In city development, soft assets are used as a tool to assist the design and management of its on-going infrastructure challenges. Soft assets are applied more broadly than hard infrastructure to maximize benefits from smart resilient city development. Compared to hard infrastructure, the measurement of soft assets is more complicated because of their unique features: (i) not inherently producing outputs; (ii) context is specific, but scalable; (iii) difficult to sell in the market (including market availability) and value is not always measured, e.g. manual, system, branding, sunk costs, etc.; (iv) combination of other assets for value creation/for sustainability (synergy); and (v) ownership of assets is not always clear, referring to a possibility of imitation (spillover). Value and impact created by these features are basically unpredictable. Haskel and Westlake (2018) summarizes the characteristics as the four Ss—scalability, spillovers, sunk cost, and synergies. Soft assets are unpredictable, often difficult to describe, and involve the 'risk of the benefit of spillover' producing unanticipated benefits and unquantifiable results which often determines how decision makers prioritize and invest in soft assets.

The creation of value from intangible assets depends on the specific city context such as policy and strategic priorities, and types of project. This process becomes even

E. Wataya and R. Shaw, *Soft Assets in Smart and Resilient City Development*, Disaster Risk Reduction, https://doi.org/10.1007/978-981-97-3373-6_3

more complex due to the nature of bureaucratic hierarchies in the public sector. Cities need to demonstrate positive performance and benefits to citizens (or taxpayers), but city bureaucratic silos often limit prioritizing soft assets to enhance service delivery. Breaking through these bureaucratic barriers is one of the major challenges in strategically incorporating intangible assets into planning, processes to create value for smart and resilient city development.

3.2 Defining Intangible Assets

3.2.1 Various Definitions, Classifications, and Measurement Approaches for Intangible Assets

Considerable previous research has involved different ways to classify and measure intangible assets. Originally, as the need to measure intangibles evolved, Kaplan and Norton (2004a, b) developed four 'core competencies' for intangible investment items: R&D, software, training and marketing (Young, 1998). These became extremely influential on later studies in this area. Table 3.1 sets out several definitions and components of intangible assets. Broadly, there are two approaches. One is to capitalize intangible assets and treat the result as a contribution to GDP. In addition to Kaplan and Norton (1996), Corrado, Hulten and Sichel (2009) also measure intangible capital by formulating how it can be incorporated into the national accounting framework. The other is a value-based approach used as a firm's performance management tool. There are different frameworks and, in this research, the latter approach is more suitable. Therefore, some of the Scorecard-type frameworks and Report-focused frameworks are briefly explained as useful in identifying intangible assets.

3.2.2 Scorecard-Type Framework

Kaplan and Norton's core components of intangibles—People, Technology, and Organizational Climate—are designed to apply to organizational resource management. Their well-known Balanced scorecard method (BSC) measures management performance with multiple perspectives combining finance and non-financial perspectives. Here, 'Financial' means the firm's outcome measures for success and supplements the metrics of 'Customer', 'Internal Process', and 'Learning and Growth' to respond to long-term shareholder value (Fig. 3.1). The goal is to facilitate and promote strategic communication among the various parties. The value of intangible assets is derived from how effectively they align with the strategic priorities of the firm (Kaplan and Norton 2004a, b). Here the level of contribution of intangibles to the firm's strategic objectives is measured rather than the costs. BSC was developed

Table 3.1 Classification of intangible categorization

Framework	Scorecard type framework				Reporting framework		Accounting-based framework	
					Knowledge management tool	Strategic management		
	Kaplan & Norton (Strategic Readiness)	Skandia (Intellectual capital)	Sveiby (Invisible capital)	Lev (Value Chain Scoreboard)	Danish Intellectual Capital Statement. Guideline	Meritum project/ Meritum Guideline	Corrado, Hulten, & Sichel (CHS) (Intangible capital)	*Corrado, Hulten, & Jona Lasinio (Public Intangibles)*
What are intangible assets/ Intellectual capitals?	Intangible assets from learning and growth objectives and value of intangible assets	Intellectual capital provides an effective instrument to manage and develop the company, as well as help strengthen competency	Intangible assets were considered in knowledge organization context and they were described as invisible capital	Intangible assets as an 'asset that is a claim to further benefits that does not have a physical or financial (a stock or a bond) embodiment' (Lev 2001)	Intellectual capital = knowledge resources 4 interrelated elements (knowledge resources = Core components) that should describe a corporation's knowledge resources	Intangible assets include three following capitals (under Intangible Resources and Activities—see below)	Intangible capital should be capitalized in national accounting system that enables to capture new economic development & change measuring economic growth	Used CHS model for categorization but broaden the concept of capital (mostly tangible) to that includes intangibles and long-lasting social assets

(continued)

Table 3.1 (continued)

Framework	Scorecard type framework				Reporting framework		Accounting-based framework	
					Knowledge management tool	Strategic management		
	Kaplan & Norton (Strategic Readiness)	Skandia (Intellectual capital)	Sveiby (Invisible capital)	Lev (Value Chain Scoreboard)	Danish Intellectual Capital Statement. Guideline	Meritum project/ Meritum Guideline	Corrado, Hulten, & Sichel (CHS) (Intangible capital)	*Corrado, Hulten, & Jona Lasinio (Public Intangibles)*
Core components/ (innovation focused) information to assess readiness	**People/Human capital** (skills, talent, employees' knowledge)	**Intellectual capital** (Product names, branded products, customers, distribution, competitors, management system, IT system, core competence, key persons, partners, etc.)	**Individual/People's competence** (Individual ability to act in various situations, including skill, education, experience, values, and social skills.)	**[S1] Internal renewal** (Research and development; Work force training and development; Organizational capital, processes) **[S1] Acquired capabilities** (Technology purchase; Spillover utilization; Capital expenditures) **[S1] Networking** (R&D alliances and joint ventures' Supplier and customer integration; Communities of practice)	**Employees**	**Human Capital** (knowledge that employees take with them when they leave the firm, including the knowledge, skill, experiences, and abilities of people)	**Computerized information** (software, databases)	**Information, Science** (software, Open data, R&D, Cultural and heritage, mineral exploration)

(continued)

Table 3.1 (continued)

Framework	Scorecard type framework				Reporting framework		Accounting-based framework	
					Knowledge management tool	Strategic management		
	Kaplan & Norton (Strategic Readiness)	Skandia (Intellectual capital)	Sveiby (Invisible capital)	Lev (Value Chain Scoreboard)	Danish Intellectual Capital Statement. Guideline	Meritum project/ Meritum Guideline	Corrado, Hulten, & Sichel (CHS) (Intangible capital)	*Corrado, Hulten, & Jona Lasinio (Public Intangibles)*
	Technology/ Information capital (database, information system, networks, technology infrastructure)	**(under Intellectual capital) Human capital**	**Internal structure** (Organization—patents, concepts, models, and computer and administrative systems)	**[S2] Intellectual Property** (Patents, trademarks, and copyrights; Licensing agreements; Coded know-how)	**Customers**	**Structural Capital** (knowledge that stays within the firm, comprising organizational routines, procedures, systems, cultures, and databases)	**Innovative property** (R&D, entertainment and artistic originals, design, mineral exploration)	**Organizational Competencies** (Brands, Organizational capital—professional and manager capital, purchased organizational services -, Function-specific human capital—employer provided training)
				[S2] Technological feasibility (Clinical tests, Food and Drug Administration approvals; Beta tests, working pilots; First mover)				

(continued)

Table 3.1 (continued)

Framework	Scorecard type framework				Reporting framework		Accounting-based framework	
					Knowledge management tool	Strategic management		
	Kaplan & Norton (Strategic Readiness)	Skandia (Intellectual capital)	Sveiby (Invisible capital)	Lev (Value Chain Scoreboard)	Danish Intellectual Capital Statement. Guideline	Meritum project/ Meritum Guideline	Corrado, Hulten, & Sichel (CHS) (Intangible capital)	*Corrado, Hulten, & Jona Lasinio (Public Intangibles)*
	Organization climate/capital (firm culture, leadership, alignment of people with firm's strategic goals, employees' ability of knowledge-sharing)	**(under Intellectual capital) Structure capital** *(under Structural capital) – Customer capital – Organizational capital*	**External structure** (Relationships with customers and suppliers, brand name, trademarks, and reputation or image)	**[S2] Internet** (Threshold traffic; Online purchases; Major Internet alliances)	**Process**	**Relational Capital** (all resources linked to the external relationships of the firm, comprising human and structural capitals, stakeholders, perceptions about the company)	**Economic competencies** (brands, organizational capital—manager capital & purchased organizational services—firm specific human capital -employer provided training)	
				[S3] Customers (Marketing alliances; Brand values; Customer churn and value; Online sales)				
				[S3] Performance (Revenues, earnings, and market share; Innovation revenues; Patent and know-how royalties; Knowledge earnings and assets)	**Technologies**			

(continued)

Table 3.1 (continued)

Framework	Scorecard type framework				Reporting framework		Accounting-based framework	
					Knowledge management tool	Strategic management		
	Kaplan & Norton (Strategic Readiness)	Skandia (Intellectual capital)	Sveiby (Invisible capital)	Lev (Value Chain Scoreboard)	Danish Intellectual Capital Statement. Guideline	Meritum project/ Meritum Guideline	Corrado, Hulten, & Sichel (CHS) (Intangible capital)	*Corrado, Hulten, & Jona Lasinio (Public Intangibles)*
				[S3] Growth prospects (Product pipeline and launch dates; Expected efficiencies and savings; Planned initiatives; Expected breakeven and cash burn rate)				
Perspectives/ Monitoring areas/ Measurement criteria/Steps	[Perspectives] – Financial matrix – Customer – Internal Process – Learning and Growth	[Perspectives/ Competencies] – Financial Focus – Customer Focus – Process Focus – Renewal & Development Focus – Human Focus	[Monitoring indicators] (i) Organizational growth/ Innovation (renewal/ utilization); (ii) Efficiency; (iii) Stability	[Measurement criteria] (i) quantitative; (ii) standardized; (iii) Empirically linked to value	[Interrelated elements in IC model] – Knowledge Narrative – Management challenge – Initiatives – Indicators	[key section of reporting] – 'Vision of firm' – 'Summary of intangible resources and activities' (in where intangible resources are classified) – 'System of indicators'		

(continued)

Table 3.1 (continued)

Framework	Scorecard type framework				Reporting framework		Accounting-based framework	
					Knowledge management tool	Strategic management		
	Kaplan & Norton (Strategic Readiness)	Skandia (Intellectual capital)	Sveiby (Invisible capital)	Lev (Value Chain Scoreboard)	Danish Intellectual Capital Statement. Guideline	Meritum project/ Meritum Guideline	Corrado, Hulten, & Sichel (CHS) (Intangible capital)	*Corrado, Hulten, & Jona Lasinio (Public Intangibles)*
Source	Kaplan and Norton (2004a)	Edvinsson and Malone (1997)	Sveiby (1997, 2010)	Lev (2001)	Guimón (2009)	Guimón (2009)	Corrado et al. (2009)	Corrado et al. (2015)

Source Wataya and Shaw (2019)

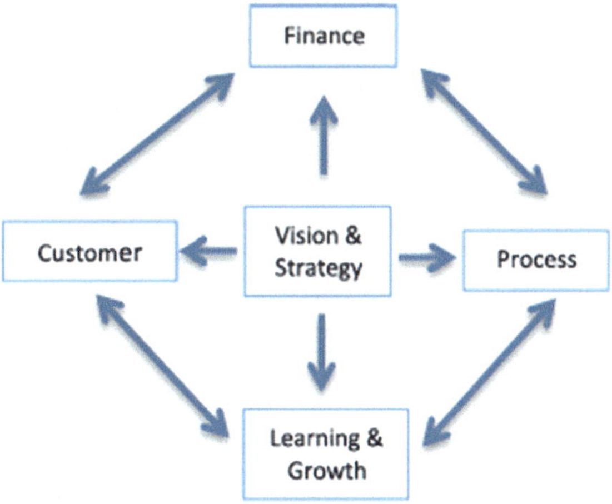

Fig. 3.1 Value creation framework (Kaplan and Norton 2004a, b) (Left)

specifically for the private sector, but it has also been applied to the public sector. This trend is reviewed and examined later in this section.

One of the value-based frameworks derived from BSC is the *Skandia Intellectual Capital Navigator* (Fig. 3.2). The Skandia Navigator used by private firms explains that intellectual capital provides an effective instrument to manage and grow a company, as well as strengthen its range of competencies. The set of competencies is similar to those of BSC, but differs in that the aim of the Skandia navigator is to measure and manage the organization's intellectual capital to enhance value by assessing and leveraging intellectual capital. The areas of focus (competencies and perspectives) are Finance, Process, Renewal & Development, Customer, and Human. Each competency has a number of indicators expressed in monetary terms (Lonnqvist 2004).

On the knowledge-focused approaches, Sveiby explains that intangible assets for a knowledge organization are 'invisible capital' (Sveiby 1997), describing people as the true agents of business and creating intangible relations and tangible products by actions and continued internal and external efforts. By contrast, the financial measurement approach is only used to assess fiscal efficiency demonstrating the need for alternative indicators beyond monetary ones. The *Intangible assets monitor* (Table 3.2) is a method to measure an organization's intangible assets by presenting three types of standards/indicators, including 'Growth/Renewal' (which indicates growth performance), 'Efficiency' (of the organization in general); and 'Stability' (the degree to which an organization can sustain its performance in a certain area is its stability/risk standard) applied to three forms of intellectual capital (Sveiby 2010).

On value creation, Lev's *Value Chain Scoreboard* approach provides a comprehensive system to evaluate the process of intellectual capital creation—value chain and innovation—and is divided into three steps, i.e. Discovery and learning; Implementation; and Commercialization (Fig. 3.3). Each step consists of three different

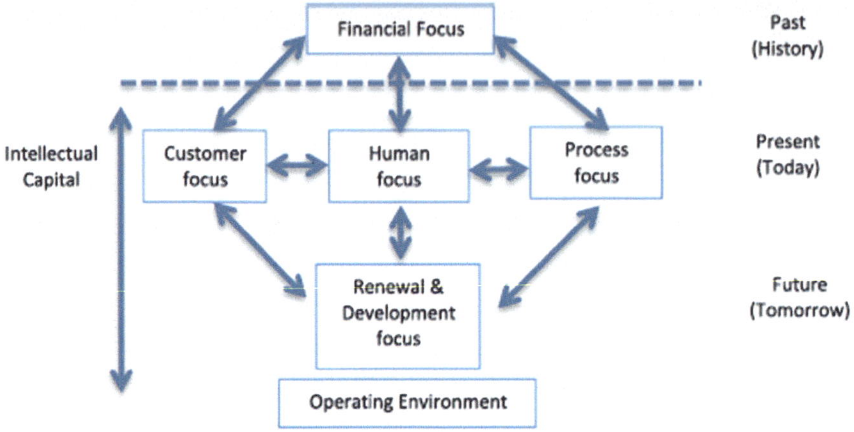

Fig. 3.2 Navigator measurement framework (Edvinsson and Malone 1997) (Right)

Table 3.2 Intangible assets monitor

The intangible assets monitor		
External structure	Internal structure	Competence
Indicators of Growth/Renewal	Indicators of Growth/Renewal	Indicators of Growth/Renewal
Indicators of efficiency	Indicators of efficiency	Indicators of efficiency
Indicators of stability	Indicators of stability	Indicators of stability

Source Sveiby (1997)

'information boxes' that are a broad cross section of economic sectors and technologies required to fulfill each of nine processes. By doing so, the measurement criteria within each of these various activities and processes aim to provide an objective result, which contributes to an organization's value. Evaluation criteria focus on several aspects, including (i) quantitative (qualitative can be supporting information to quantitative aspects); (ii) standardized (for easy comparison across firms); and (iii) Empirically linked to value (Lev 2001).

3.2.3 *Key Lessons*

Different frameworks underlie the alternative approaches and methods of organizational resource management. Some can be applied to the national or local governments administration involved in smart, resilient city development. Kaplan's BSC emphasizes internal process improvement within organizations. The Skandia Intellectual Capital Navigator focuses on intellectual capital, emphasizing knowledge creation, sharing, and utilization to enhance structural capital. Sveiby's intangible

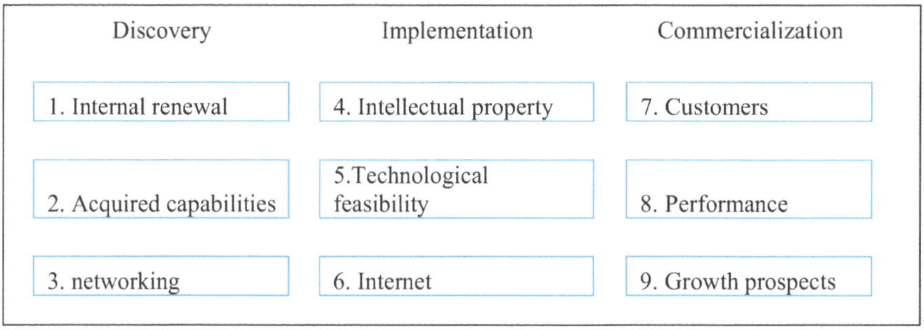

Fig. 3.3 The value chain scoreboard. *Source* Lev (2001)

assets monitor views intangible assets as a competitive advantage, with processes related to knowledge and innovation contributing to their development. Lastly, Lev emphasizes activities and processes within the value chain to illustrate the creation and contribution of intangible assets for value.

These various frameworks identified common and distinct areas relevant to the potential value creation model in smart resilient city development initiatives, providing insights for framework modification. First, private firms assign more value investing in soft assets to develop and nurture their firms. This could not be done without clear objectives and strategies, including results firms want to achieve. Second, the organizational resources management system in firms' soft assets is designed for the effectiveness of client service delivery through their products and services that benefit clients. Third, although providing solutions for clients and effective service add to a firm's value, this is closely linked to strengthening a firm's human capital and accumulated technical and operational knowledge. Therefore, the firm's products or services alone are not objectives, but rather ways to accomplish a firm's business model. Fourth, achieving these desired outcomes cannot be produced in a siloed working environment. A process that aims to fulfill different areas of expertise and functions requires broad cross-sectional collaboration to provide objective results. These efforts encourage activities to be scaled-up to meet higher outcomes. Lastly, open communications across the varied levels of staff in a firm are important. This facilitates an open environment with a balanced top-down and bottom-up culture. The process of creating value needs involvement of all levels of stakeholders participating in an iterative process.

3.3 Types of Soft Assets Reviewed in This Research

Based on the brief review of different objectives for the classification and measurement of intangible assets, this research discusses several leading studies by Kaplan and Norton (strategic readiness), Sveiby (invisible capital), and Edvinsson and Malone (intellectual capital) for identifying soft asset capital (Table 3.3).

Table 3.3 Major models that were referred to this research

	What are intangible assets/ Intellectual capitals?	Core components of intangible assets
Kaplan and Norton (2004a, b) *(Strategic Readiness)*	Intangible assets from learning and growth objectives and value of intangible assets	People/Human capital; Technology/Information capital; Organization climate/ capital
Skandia (Intellectual capital) by Edvinsson and Malone (1997)	Intellectual capital provides an effective instrument to manage and develop the company, as well help strengthen competency	Intellectual capital; (Under Intellectual capital) Human capital; Structure capital;
Sveiby (2010) *(Invisible capital)*	Intangible assets were considered in knowledge organization context and they were described as invisible capital	Individual/People's competence; Internal structure; External structure

Source Wataya (2023)

Kaplan and Norton identified intangible assets from learning and growth objectives and emphasized the value of intangible assets (Kaplan and Norton 2004a, b); Sveiby defined intangible assets in the context of knowledge organization and described as invisible capital (Sveiby 2010); while Edvinsson and Malone (Edvinsson and Malone 1997), using the Skandia Intellectual Capital Navigator measurement framework (Lonnqvist 2004), identified intellectual capital that provides an effective instrument to manage and develop corporate activity. Corporate soft assets are those assets which are developed and sustained by learning for growth, serve as engines of development within a knowledge organization environment and are able to help strengthen corporate expertise. As this analysis shows, corporate practices are reliable precedents for selecting appropriate soft assets within the urban environment.

To define soft asset capital for this research, Table 3.4 shows the broader groups of soft asset components (modified by individual context).

These are broadly similar to the intangible asset components that are reviewed in the earlier section. In the research, sustainable urban development for smart resilient cities is captured by strategic integration of soft assets in public sector development initiatives to achieve desired outcomes. The definition of soft assets in this context is capital components that develop an enabling environment to build and maintain the key value elements which constitute smart, resilient city development.

While each of the soft asset categories is linked, they are broadly divided into several groups, related to organizational capital: i.e. (a) Institutional assets; (b) Framework and Systems; (c) Social capital-related, i.e. Social assets; (d) Intellectual products; and (e) Computerized products and information, i.e. Information and knowledge asset-related. The aim of this preliminary categorization is to streamline key soft asset elements to identify the effective integration of soft assets and exploring potential methods of assessing outcomes and impact. Due to the intangible nature

Table 3.4 Broader set of soft assets classifications

Classification	Hard assets	Soft assets	
Sub-classification	Hard assets	Soft assets	Medium assets
Type of assets	(a) Assets – Physical infrastructure	(a) Institutional assets (organization capital) – Individual capital, knowledge (function-specific) – Institutional capital, knowledge, professional technical knowledge and experiences, management capacity, – Training (both for individuals and institutional contexts) – Learning and growth capacity – Relationship with external stakeholders – Institutional credibility, reputation	(e) Computerized products and information (information and knowledge-related capital) – Apps – Database (Information and/or applications that can be integrated into physical infrastructure)
		(b) Framework and systems (organization capital) – Policy, strategies, plan,	
		(c) Social assets (social capital) – Social system – Community network – Social norms, value	
		(d) Intellectual product (information and knowledge-related capital) R&D, Reports	

Source Wataya and Shaw (2019)

of soft assets, impact is more accurately assessed through a combination of qualitative and quantitative analyses, but recognizing the overall importance of quantitative analysis.

The classification can be further subdivided using different levels of capital as enabling factors, e.g. individual capital or people, human capital, system, or process capital. An example of an enabling factor is where local authorities, as the main provider of public works, strengthen the governance system, including policy-making and planning, to ensure the best services are provided to their citizens with innovative solutions supported by soft assets. To build a solid foundation to accomplish city goals and responsibilities, the implementation of public works requires dealing with a range of issues, including political, analytical, planning, systems, and accountability.

Regarding 'Computerized products and information', ICT-linked hard infrastructure is becoming a more important factor in recent city development. Therefore, soft assets are divided into 'Soft assets' and 'Medium assets' as elaborated in Table 3.4. Data, apps, and other related technologies are included in this category. A useful distinction is to classify the assets by the 'degree of soft assets' to understand the type and nature of soft assets. This means, for example, technical soft assets, such as computerized products, have an obvious value, while the more indescribable soft assets, such as human, institutional knowledge, and capital, are more difficult to quantify. The important point is that these assets, although difficult to assess, are nevertheless a primary source of unique value creation in smart, resilient city development.

Based on the broader set of soft asset classification, and also taking into consideration the enabling factors for soft assets, Table 3.5 shows the list of potential soft assets, which are further refined based on the information in Table 3.4, and these refined assets are utilized for this research. Social capital is often understood and identified from a community-based or bottom-up perspective. Hence two levels of descriptions were also included.

Table 3.5 Five soft asset capitals

Soft asset capitals	Description
People/ Human capital	• Technical, operational individual capacity (Knowledge that employees take with them when they leave the firm, including the knowledge, skills, experiences, and abilities of people)
Institutional capital	• Legislation, policies, strategies, standards, frameworks, systems, mechanisms, modality, and other relevant items. They are backbone of smart city development • Processes, procedures, data, systems, culture, designs, and knowledge (knowledge retention to execute tasks and responsibilities
External capital	• Relationships with stakeholders such as citizens, donors/financial sources, private sector, academia, LGU, SPV *(depends on your roles)* • Network or any form of crating benefits to stakeholders for promoting relationships
Technology, information capital	• Computerized products, databases, information systems, networks, technology infrastructure, also covering a wide meaning of the use of technology for smart city development
Social	• Community rules, regulations, framework, community representative groups (e.g. contribute to LUG's planning/consultation process) • Local level organization, values (trust, solidarity, reciprocity), social norms, behaviors, attitudes (strength, number, quality of relationship among people in a society)

Source Wataya and Shaw (2022)

3.4 The Benefit and Challenges of Soft Assets for Smart, Resilient City Development

Soft assets provide value-added creation opportunities and the process should be an integral part of resilient city development. As examined in the previous chapter, 'resilience is the capacity of a social-ecological system to absorb or withstand perturbations and other stressors such that the system remains within the same regime, essentially maintaining its structure and functions'. It describes the degree to which the system is capable of self-organization, learning, and adaptation (Resilience Alliance).

From the sociological perspective, many key soft assets are focused on people and human and institutional capital as enabling factors to create robust community (social) systems and organizational structures. UNISDR defines a resilient city as a city capable of withstanding and absorbing the impact of hazards, shocks, and stresses through adaptation or transformation, in order to guarantee long-term sustainability, as well as its basic city functions, characteristics, and structures (UNISDR 2012).

Adaptive capacity is also developed through a repetitive learning process and growth from lessons from past experience. Norris et al. (2008) emphasized: 'community resilience emerges from a set of capacities—community resilience from a set of networked adaptive capacities. And resilience rests on both the resources themselves as well as the dynamic attribution of those resources'. Therefore, adaptive capacities consist of a set of *networked resources*, i.e. economic development, social capital, information, communication, and community competencies, that are supported by soft assets.

Soft assets increase resiliency at different community levels within a city. However, they are intangible, difficult to quantify, with undefinable ownership and are unpredictable due to their intangible characteristics. As explained by Haskell and Westlake (2018), the 'risk of the benefit of spillover' makes soft assets difficult as an investment target by city decision makers. There is a challenge for both the private and public sectors to show clear evidence that investment in soft assets creates positive outcomes and impact for their stakeholders and beneficiaries. In addition, public sector bodies have made less effort to value and deliberately incorporate and value intangible assets because of the multiple objectives of the non-financial nature and intangible final products (Serrano Cinca et al. 2001). Given this situation, soft assets in the city context are less likely to be a tangible target for public investment. To effectively address citizens' needs and provide comprehensive functional city services and build smart, resilient cities, city administrators need to achieve a balance between the investment in hard and soft assets.

3.5 A Framework for Value Creation for Smart, Resilient City Development

Earlier in this chapter, the various management tools associated with soft assets used by the private sector were reviewed. Those tools not only helped firms better measure management performance, but also identify and understand which soft assets are required in specific cases. This methodology can equally be applied to the public sector. This section briefly reviews the BSC value creation model and how it can be applied to smart, resilient city development.

In the value creation model, a city has its own 'mission to accomplish' while a firm has its business and profit enhancing 'strategy to achieve' their objectives (Kaplan and Norton 2004a, b). A major goal of a city is to efficiently provide quality services to satisfy their citizens' needs. The 'Customers' perspective in the private sector is not equivalent to that of public sector citizens who are both beneficiaries consuming services as well as stakeholders and taxpayers. Citizens monitor and know whether their city's service delivery is fair, transparent, effective, and consistent. Kaplan and Bower (1999) explain that public organizations have three significant objectives to accomplish their mission, i.e. Cost incurred, Value created, and Legitimizing support (Fig. 3.4). These are all related to the differences mentioned above. Cost includes the public organization's direct expenses and social costs to be covered by citizens and service providers. Minimizing these costs and optimizing delivery of services to meet the needs of citizens is critical. Value creation is difficult to measure, but Kaplan explained that organizations can identify outputs by the balanced scorecard and benefits accruing based on their outputs and inputs. *Legitimizing support* is associated with funds (= financier/donor/taxpayer) required to provide stable and quality services. To do so, the actions of cities need to be justified and demonstrate accountability.

For this research, three higher level objectives reflecting the characteristics of public organizations are useful to identify smart, resilient city development. The concept of competitiveness in the private sector equates to 'value creation' in smart, resilient city development. A long-term strategy which sets a clear vision is indispensable to improve community well-being, including basic service provision, safeguarding the environment, and creating job opportunities for economic development.

To achieve these significant higher level objectives, it is necessary for cities to understand the needs of citizens and understand how to improve service delivery to meet their needs. Also it is important to identify key city domains or areas where soft assets can play an important role to create value. This is a natural convergence of a city's policies and plans. This is because a city's higher level objectives are achieved by projects and initiatives designed for citizens (= taxpayer) and other stakeholders. Furthermore the quality of projects and activities to contribute to these three objectives should be developed within a learning and growth process that underpin this value creation framework. One example is the value associated with the social benefits provided by social infrastructure (e.g. hospital, school, housing, and others).

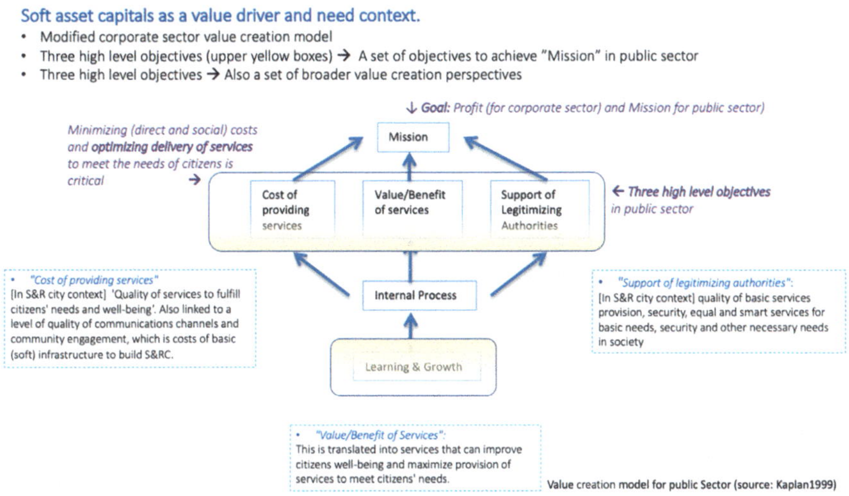

Fig. 3.4 Value creation framework for public organization. *Source* Kaplan and Bower (1999)

Corrado et al. (2015) stated that this social benefit is the spillover or an externality that results from citizens consuming the social goods and services. Creating and delivering services are supported by hard infrastructure, but greatly strengthened by soft assets, which are a dominant source of innovative solutions. Cities, therefore, have to make deliberate choices about policies and strategies to deliver optimal services to their citizens. Their projects, in most cases, generate multi-dimensional, multi-sectoral results and impact.

Challenges in applying the BSC framework to public organizations

There are some challenges in applying the BSC model to public organizations. As shown in Fig. 3.4, BSC principles applied to public sector organizations need adjustments in design and implementation due to its unique institutional and structural environment (Kaplan 2001 in Gadenne and Sharma (2009)). Some challenging areas emerge in articulating city mission strategies such as the difficulty to select and incorporate the key BSC performance indicators due to inconsistencies with existing medium- to long-term policy or outline of administrative reform. This also links to prioritizing the introduction of performance measurement systems in the public sector environment rather than key performance indicators in the four BSC perspectives. Gadenne and Sharma (2009) mentioned that BSC implementation is challenging due to planning, deployment, review, motivation, leadership, resources, communication, measurement, and date collection techniques, as well as the role of knowledge management. BSC is a powerful tool to implement strategy which embraces a consideration of intangible assets and tangible assets. The aim of this research is not to apply BSC, but apply elements of its value creation system to the method of identifying the consideration and use of soft assets.

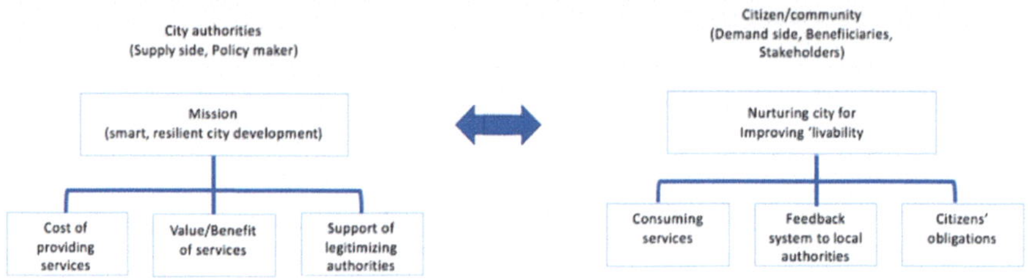

Fig. 3.5 Soft assets co-value creation in smart, resilient city development (city authorities and citizen/community). *Source* Created by the author based on Kaplan and Bower (1999)

Co-value creation model

Considering that the development of smart, resilient cities is achieved at several levels, there is always a two-way interaction between the city authority (as a service provider) and citizens (as consumers, beneficiaries, and stakeholders). This aspect is set out in Fig. 3.4 which indicates the city side of the value creation model. But the value-added of soft assets is created at both ends of the spectrum and their processes are invariably linked, creating synergies for value creation. Figure 3.5 shows a preliminary type of value creation model that is developed in this research.

Using the idea of co-value creation, the framework above combines two sides of the equation—providers (cities) and beneficiaries (citizens/community). In reviewing Fig. 3.5 above, three key objectives can be developed at the city level:

- 'Cost of providing services' that is closely linked to 'Legitimizing support' in terms of a source of funding (i.e. tax payer's money). In the smart, resilient city context, this can translate into 'quality of services to fulfill citizens' needs and well-being'. The foundation of these services is also linked to a level of quality of communications channels and community engagement, which is costs of basic (soft) infrastructure to build smart city.
- 'Value/Benefit of Services': This is translated into services that can improve citizens' well-being and maximize provision of services to meet citizens' needs.
- 'Support of legitimizing authorities': This means city authorities or service providers need to satisfy citizens with their services. Since they are both clients, as well as a source of financial resources (i.e. tax money) for provision of services, cities have to constantly factor in effective planning for quality of basic services provision, security, and other necessary social needs.

On the citizen and community spectrum, three corresponding objectives for reciprocal relations are listed in Fig. 3.5:

- Consuming services: Citizens consume services provided by cities, which in turn corresponds to the city's cost of providing these services. The consumption represents the utilization of resources allocated through taxes to benefit the community.

- Feedback system to cities: Citizens play a crucial role in providing feedback to city authorities regarding the quality, effectiveness, and value of the services received. This feedback helps the city authority to make improvements and better align their services to meet the needs and expectation of the community. The community needs to be able to engage with, raise issues, and identify problems with city authorities, to promote accountability.
- Citizen's obligations: This corresponds to 'Support of legitimizing authorities' where citizens have an obligation to pay taxes, which is the main source of revenue for cities.

The concept here is that the value creation process occurs on both sides—city administration and citizens—also within the city as a whole. The assumption is that there is a two-layered horizontal and vertical reciprocal value creation process. The value creation framework creates a foundation for quality of life, benefiting from 'smartness' by using technology, and strengthening the city's capacity for resilience. Identifying embedded soft assets and aspects of value creation will be examined in the next section.

3.6 Defining Nine Soft Asset Value Elements Essential for Smart, Resilient Cities

Identifying soft assets integration and defining value creation requires value creation element or a context. To define the key soft asset value elements in smart, resilient cities for this research, several key smart city pillars were examined.

Giffinger et al. (2007) ranked European medium-size cities and identified six smart city pillars. This is an appropriate starting point to examine which elements are more significant for soft assets value creation. While there are different ways to view smart city characteristics, most fall into these categories. On city resilience, ICLEI's report on connecting green recovery and resilience in cities (ICLEI 2021a), listed in the Resilient Recovery Framework four key factors that bring multiple benefits to communities, namely (a) equity-centered (empowering multiple stakeholders); (b) risk-awareness (planning for recovery to leverage date and technology); (c) system-enabled (recovery intervention design in terms of both multiple benefits and nexus between different systems); and (d) climate-focused (fossil-fuel to renewable energy based recovery planning) (ICLEI 2021a).

Considering that key pillars or themes in various indicators have greater synergies with the above dimensions, they can greatly assist identification of soft asset value elements in smart, resilient cities. For this research, key elements are selected by referring to the references above and the higher objectives inherent in the public sector value creation framework. There are different classifications for key smart city pillars, but Giffinger et al. (2007) created a comprehensive model that can be adapted for this research. Major aspects of resilience should also be included. Giffinger et al. (2007) further subdivided each of the six smart city characteristics with assigned sub-factors, then created 31 sub-factors with 74 indicators to cover important city elements. This is significant as it involves city functions, activities, community engagement, communications, and factors related to social capital.

Table 3.6 shows a 'comparison between different dimensions of smart cities' and Fig. 3.6 shows nine value elements essential for smart, resilient cities.

In terms of value creation, three higher objectives provide guidance to identifying key smart, resilient city development. These previously mentioned three objectives can also be useful for identifying the context for soft assets in this research as follows:

(i) **Cost of providing services**: Cost effectiveness of providing basic services is important in smart, resilient cities. Thus, quality of services to fulfill the citizens' needs and well-being is achieved by better resource allocation. It relates to a system or platform for quality of communications and channel of community engagement to ensure dialogue and feedback from the ground up.

(ii) **Value/Benefit of services**: This also relates to innovative approaches to create new value and innovation. This can be translated into maximum level of service delivery that can improve citizen's well-being, with smarter services and technology to meet citizens' needs.

(iii) **Support of legitimizing authorities**: Under this objective, citizens pay taxes in exchange for receiving essential services from the city. Citizens are both

Table 3.6 Comparison between different dimensions of smart cities

Giffinger et al. (2007)	ICLEI	Kaplan	This research
Smart city 6 dimensions: Ranking for European medium-sized cities	Green recovery and Resilient cities	Value creation model for public organizations—3 high-level objectives	Soft asset value elements
Smart economy	Equity-centered	Cost of providing services	Resource management
Smart people	Risk awareness	Value/benefit of services	Mobility (Accessibility)
Smart governance	System enabled	Supporting of legitimizing authorities	Health, Well-being
Smart mobility	Climate-focused		Equal access to services
Smart environment			Smarter services via technology
Smart living			Community engagement
			Quality communications
			Security
			Resilience

Source Giffinger et al. (2007), ICLEI (2021a) with author's edition

Fig. 3.6 Nine value elements essential for smart, resilient cities. *Source* Wataya and Shaw (2019), Wataya (2023)

taxpayers and customers of the city. Provision of services that keep citizens safe and enhance security is key. So, the quality of basic service provision, security, equal access, level of smartness, and resilience are basic social needs.

Based on the framework examined above and factoring in soft asset capitals, the nine soft asset value elements are selected as follows (Wataya and Shaw 2022): (i) resource management: (ii) accessibility (mobility); (iii) health and well-being; (iv) equal access to services; (v) smarter services; (vi) community engagement; (vii) quality communication; (viii) security; and (ix) resilience (Fig. 3.6). These nine value elements are unique and essential to this research, but are also quite broad in scope. If soft assets are strategically and successfully embedded, they create value for smart and resilient city. The purpose is not to list various areas for quantifying results, but point to anticipated value creation clusters to identify the level of soft assets integration. The nine elements cover the key clusters.

Resource management aims to optimize limited human and natural resources essential to build a smart and resilient city. This element covers not only the effective use of resources, but focuses on efficient systems. It reflects strategic readiness, planning, competencies in resource management planning, benefits to stakeholders, and use of technology to optimize outcomes.

Accessibility (mobility) is designed to achieve reduced automobile dependence, providing alternative modes of transportation. It reflects the capacity to establish a holistic urban development plan. It also addresses availability of cross-sectoral planning, provision of a citizen-friendly public transport system, implementing an IT-led system, and facilitating the engagement of citizens in planning.

Health, well-being emphasizing building an environmentally sound (green), healthy environment and community for citizens; a healthy living space and a positive impact on human health by creating public gathering places, improving accessibility to services, and incentivizing the community to support and participate in improving health-related initiatives.

Equal access to services is designed to maintain social stability and securing support from citizens, stakeholders, and beneficiaries. It focuses on satisfying basic community service needs and measures to mitigate risks, create sustainable financial options, strategic readiness, improving citizens' tracking process for accessing services, improving IT literacy, and facilitating better levels of citizen engagement and decision-making processes.

Smarter services via technology provides access to effective services that meet the needs of citizens and improve service quality using smart infrastructure. It concentrates on developing a comprehensive vision and strategy across sectors, readiness, availability of plans and systems to improve citizens' technical literacy, establishing efficient communication channels and facilitating improved levels of citizen engagement and decision-making.

Community engagement underlines the improvement of quality of life by strengthening the community's capability to communicate and become involved. It focuses on developing effective community engagement plans, including public awareness raising activities, strategic readiness, availability of support for community-based awareness raising efforts, strengthening community engagement in public events, providing accessible technology and IT systems for an inclusive community, improving levels of citizen engagement, and contributions to strengthen social values.

Quality communications stresses the need to improve communications between local government and communities to contribute to building a sustainable and resilient smart city. It seeks to establish a citizen–government communications channel, streamlined coordinating mechanisms between the city and stakeholders, regularly conducting feedback on the provision of city services, a feedback collection system, a collective/transparent decision-making process, and a method of accountability of leaders.

Security emphasizes providing a safe and secure environment to protect citizens' daily life and improving smart city resiliency against unexpected external shocks and disasters.

Resilience refers to making a city stronger against external shocks and risks, preparing it for short-term hazards and mid- to long-term adaptability. The resilience element refers to the Climate Disaster Resilience Index methodology (Joerin et al. 2014) as it has already defined the resilience of communities in a city in relation to climate-related disasters. In addition, this research extends aspects of resilience beyond natural disaster risk management. It covers the need for several key elements: a crisis zone management framework, natural resource management, knowledge management and dissemination, mainstreaming of disaster risk reduction, climate-change adaptation, good governance, land use planning, effective transport, clean energy, institutional collaboration with other organizations and stakeholders during a disaster, effectiveness of early warning systems, internet access, community disaster preparedness, education and awareness and strengthening social capital (Wataya and Shaw 2019; Wataya 2023).

All nine elements are considered the main pillars which underpin the development of soft asset consideration indicators. They attempt to set benchmarks to enable a city to improve its safety plan for better risk management, to develop a framework for ensuring the welfare of citizens in times of crisis, developing community security services, enabling police departments and law enforcement to establish a safe and secure city, provide quality of data management systems, collective/transparent decision-making processes, accountability of leaders, and strengthening social values. A summary of the definitions of the nine value elements in this research is shown in Table 3.7.

Qualitative framework for identifying embedded soft assets in nine value elements

To identify and analyze the process of value creation within city development activities, a soft asset identification tool with 45 items was developed based on nine elements and five soft asset capital components (Table 3.5). The nine value elements are closely related to each other and there is some duplication across the elements. For example, the 'smarter services via technology' element is closely interlinked to 'security' and 'resilience'; also 'quality communications' and 'community engagement' embody similar activities. This comprises two factors: (i) soft asset value elements in smart and resilient cities, and (ii) five soft asset capital components. Together with the five soft asset capital components, the proposed evaluation tool comprises 45 items ($9 \times 5 = 45$). The objective of the tool is to show which competencies can be employed by cities to deliver smarter quality services.

Soft asset capital, although obscure and granular, needs context to generate value through synergies, spill-over, and scaling up. Specific values are created in planning initiatives or projects, but the tool is not designed to evaluate performance at project level, but to assess the utilization of soft assets to specifically help meet the vision and objectives of a smart, resilient city. Therefore, unlike other mutually exclusive indicators, the proposed tool is not purely for quantitative assessment. It extracts aspects of soft asset consideration from information in smart city proposals. This implies that the qualitative framework with 45 items shown in Table 3.8 may overlap with each other at some level depending on the type of soft asset capital and the

Table 3.7 Definitions of the nine value elements essential for smart, resilient cities

Resource management	To optimize limited human and natural resources for building a smart and resilient city
Accessibility (mobility)	To facilitate reduced automobile dependence, providing alternative modes of transportation
Health well-being	To build an environmentally sound (green), healthy environment for citizens
Equal access to services	To maintain social stability
Smarter services via technology	To provide access to effective and efficient services that meet the needs of citizens and improve service quality through the use of smart infrastructure
Community engagement	To improve quality of life by strengthening the community's capability and nurture the city by citizens' close involvement over a long time
Quality communications	To improve communications between local government and communities to build a sustainable and resilient smart city
Security	To provide a stable, secure environment in citizens' daily life and improve smart city resiliency against unexpected external shocks and damage
Resilience	To make a city stronger against external shocks and risks, prepare it for short-term hazards and mid- to long-term adaptability

Source Wataya and Shaw (2022), Wataya (2023)

value elements involved. The focus on qualitative aspects rather than numeric data is a limitation of this type of tool. There are some additional aspects about this qualitative tool which are relevant as follows: (i) some overlapping is inevitable due to the qualitative nature of the tool; (ii) 'smarter service via technology' (value element); and 'technology and information' capital (soft asset capital) which may overlap to some extent, but there is a difference. Both are related to technology, but the value element exists where soft asset capital can yield value; (iii) there may be some overlap at two different levels within the same elements (e.g. 'use land pooling for real estate creation in eastern periphery of city' under 'resource management' where both are connected by 'people/human capital/ and 'institutional capital'). Another case is comparing different elements (e.g. Transit Oriented Development for metro root development' under 'accessibility (mobility)' and 'health, well-being' are both connected with 'institutional capital' and 'external capital'). Greater detail on these overlapping characteristics will be examined in Chap. 5. Although each element is not exclusive or all encompassing and may overlap, it is nevertheless possible to identify the degree of soft asset incorporation. However, if the five soft assets and the nine value elements are further subdivided into more detailed sub-categories, it reduces the possibility of overlapping.

Table 3.8 Qualitative framework to identify an adaptation of soft asset incorporation

Soft asset value elements	Soft asset capitals
Resource management	People/Human capital (competency in resource management related policy, strategy, planning development, air quality, biodiversity, water conservation, energy efficiency, SWM, etc.) Institutional capital (strategy, planning, approach to mainstream resource efficiency in city development policy and strategy) External capital (coordinating mechanisms between LGU and stakeholders to embrace unique local character, benefits to stakeholders (citizens) for promoting resource management) Technology (information capital) (promotion of automation-based smart resource management systems and technology in public service provision) Social (structural) (level of contribution to strengthening collective/ transparent decision-making processes) Social (cognitive) (facilitating level of participation in policy/planning processes)
Accessibility (Mobility)	People/Human capital (competency in establishing holistic urban development plans) Institutional capital (availability of cross-sectoral coordinated land-use planning (and readiness for implementation)) External capital (provision of citizen friendly public transport services; benefits to stakeholders (citizens) for promoting accessibility) Technology (information capital) (developing data management tools, digital technology for better resource management) Social (structural) (level of contribution to strengthening collective/ transparent decision-making processes) Social (cognitive) (facilitating level of participation in policy/planning processes)
Health well-being	People/Human capital (competency in establishing a shared vision of a healthy community for residents; better and healthy living places, good green spaces; equal and inclusive society) Institutional capital (creating policies and a system for making health choices easy for citizens; better and healthy living places; equal and inclusive society) External capital (making a direct, positive impact on human health by creating public gathering places; benefits to stakeholders (citizens) for improving health/well-being) Technology (information capital) (developing data management tools, digital technology for improving accessibility to healthcare service facilities, healthy living environment) Social (structural) (level of contribution to strengthening collective/ transparent decision-making processes) Social (cognitive) (facilitating level of participation in policy/planning process, making values (trust, solidarity, reciprocity) stronger in policy and systems development processes)

(continued)

Table 3.8 (continued)

Soft asset value elements	Soft asset capitals
Equal access to services	People/Human capital (competency in identifying basic community service needs and implementing countermeasures for risks) Institutional capital (availability of sustainable financial strategies to ensure equal access to services) External capital (development or update of systems for monitoring and integration to identify key challenges and track progress over time; benefits to stakeholders (citizens) for improving equal access to services) Technology (information capital) (reduction in the number of citizens without IT systems (internet connectivity)) Social (structural) (level of contribution to strengthening collective/transparent decision-making processes) Social (cognitive) (facilitating the level of participation in policy/planning processes (Citizens can easily engage in public service monitoring to provide feedback on services provided by LGU))
Smarter services via technology	People/Human capital (competency of developing holistic visions/strategies across sectors) Institutional capital (readiness, availability of the plan/system to improve citizens' technical literacy) External capital (establishing efficient communication channels within an organization; benefits to stakeholders (citizens) for improving smarter services) Technology (information capital) (readiness, availability of system for easy access to LGU services and data, tech-linked services, smarter public services) Social (structural) (level of contribution for strengthening collective/transparent decision-making processes) Social (cognitive) (facilitating levels of participation in policy/planning processes)
Community engagement	People/Human capital (competency of developing effective community engagement plans, including public awareness raising activities) Institutional capital (readiness, availability of support for community-based awareness raising efforts (for building a better city)) External capital (planning community activities for strengthening community engagement in public events; benefits to stakeholders (citizens) for improving community engagement) Technology (information capital) (readiness of feasible technology and IT systems for elderly and vulnerable citizens for an inclusive society) Social (structural) (level of contribution to strengthening collective/transparent decision-making processes) Social (cognitive) (values (trust, solidarity, reciprocity))

(continued)

Table 3.8 (continued)

Soft asset value elements	Soft asset capitals
Quality communications	People/Human capital (competency of capability to establish citizen–government communication channels) Institutional capital (creating streamlined coordinating mechanisms between LGU, the private sector, and the community for effective communication) External capital (regularly conducting feedback on city service provision quality from citizens; benefits to stakeholders (citizens) for improving quality communications) Technology (information capital) (availability of feedback collecting systems and tools for citizen) Social (structural) (collective/transparent decision-making process, accountability of leaders, practices of collective action and responsibility) Social (cognitive) (level of decentralization, facilitating level of participation in policy process)
Security	People/Human capital (competency of capability to improve city safety plans for better city risk management) Institutional capital (developing a framework for ensuring the welfare of citizens in times of crisis) External capital (developing community security services, availability of enabling police departments and law enforcement for safe cities; benefits to stakeholders (citizens) for improving security) Technology (information capital) (availability of quality of data management systems/frameworks for citizen to access, planning security measures, handling options) Social (structural) (local level organization, collective/transparent decision-making processes, accountability of leaders, practices of collective action and responsibility) Social (cognitive) (support for strengthening local values (trust, solidarity, reciprocity), social norms, behavior for improving local security)
Resilience	People/Human capital (effectiveness of a zone's crisis management framework, knowledge dissemination and management, natural resource management) Institutional capital (mainstreaming of disaster risk reduction and climate change adaptation, good governance, land use plans, green icon-related planning, transport, clean energy) External capital (institutional collaboration with other organizations and stakeholders during a disaster) Technology (information capital) (effectiveness of early warning systems, access to the internet) Social (structural) (community preparedness during a disaster, education, and awareness) Social (cognitive) (social capital, localism)

Source Wataya (2023), Wataya and Shaw (2022)

By using the proposed tool, the information selected from the materials is examined and assessed for the level and strength of connectivity between selected actions or initiatives associated with soft asset capital components in Chap. 5.

References

Corrado C, Hulten C, Sichel D (2009) Intangible capital and U.S. economic growth. Rev Income Wealth 55(3):661–685

Corrado C, Haskel J, Jona-Lasinio C (2015) Public intangibles: The public sector and economic growth in the SNA, SPINTAN working paper series no. 1 Instituto Valenciano de Investigaciones Economicas, S.A. 32

Edvinsson L, Malone M (1997) Intellectual capital: The proven way to establish your company's real value by measuring its hidden brain power. Piatkus Book, London, p 225

Gadenne D, Sharma B (2009) Balanced scorecard implementation in the public sector: lessons learnt in a large local government authority, pp 1–25. In: Accounting and finance association of Australian and New Zealand (AFAANZ) conference (Adelaide, Australia 5–7 July 2009) https://research.usc.edu.au/discovery/fulldisplay/alma99449692702621/61USC_INST:ResearchRepository Document Type: Published Version

Giffinger R, Fertner C, Kramar H, Kalasek R, Pichler-Milanovic N, Meijers E (2007) Smart cities, ranking of European medium-sized cities. Final report, October 2007, Delft University of Technology. https://www.smart-cities.eu/download/smart_cities_final_report.pdf

Guimón J (2009) MERITUM and Danish guidelines for reporting on intangibles: a comparative study. ICFAI Univ J Account Res VIII(2):17–29

Haskel J, Westlake S (2018) Capitalism without capital: the rise of the intangible economy. Princeton University Press, Princeton, New Jersey, USA, pp 221–222

ICLEI (2021a) 'Resilient Cities Action Package 2021 (ReCAP 21)', ICLEI—Local Governments for Sustainability, Bonn Germany. https://resilientcitiesnetwork.org/downloadable_resources/Programs/ReCAP21_Guide.pdf

Joerin J, Shaw R, Takeuchi Y, Krishnamurthy R (2014) The adoption of a climate disaster resilience index in Chennai, India. Disasters 38(3):540–556. https://doi.org/10.1111/disa.12058

Kaplan RS, Bower M (1999) The balanced scorecard for public-sector organizations, balanced scorecard report—insight, experience & ideas for strategy-focused organization. Harvard Business School Publishing, Massachusetts, Boston, MA, USA

Kaplan RS, Norton DP (1996) The balanced scorecard: translating strategy into action. Harvard Business School Publishing, Boston, MA, USA

Kaplan RS, Norton DP (2004a) Measuring the strategic readiness of intangible assets. Harv Bus Rev https://hbr.org/2004/02/measuring-the-strategic-readiness-of-intangible-assets

Kaplan RS, Norton DP (2004b) Strategy maps: converting intangible assets into tangible outcomes. Harvard Business School Publishing, Boston, MA, USA

Lev B (2001) Intangibles: Management, measurement, and reporting. The Brookings Institute Press, Washington, D.C., USA

Lonnqvist A (2004) Measurement of intangible success factors: case studies on the design, implementation and use of measures. Tampere University of Technology Publication 485, Tampere, Finland

Norris FH, Stevens SP, Pfefferbaum B, Wyche KF, Pfefferbaum RL (2008) Community resilience as a metaphor, theory, set of capacities, and strategy for disaster readiness. Am J Community Psychol 41(1–2):127–150. https://doi.org/10.1007/s10464-007-9156-6

Resilience Alliance, 'Resilience', Resilience Alliance, 'Resilience' (2015).

Serrano Cinca C, Mar Molinero C, Bossi Queiroz A (2001) An approach to the measurement of intangible assets in public sector using scaling techniques. Discussion Papers in Accounting and Finance, University of Southampton, Number AF01-4, ISSN 1356-3548

Sveiby KE (1997) The intangible assets monitor. J Human Res Costing Account, 2(1):73–97

Sveiby KE (2010) Methods for measuring intangible assets. https://www.sveiby.com/files/pdf/intangiblemethods.pdf

The United Nations Office for Disaster Risk Reduction. Making Cities Resilient Report 2012, UNISDR (2012)

Wataya E, Shaw R (2019) Measuring the value and the role of soft assets in smart city development. Cities 94:106–115. https://rajibshaw.org/wpRS/wp-content/uploads/2019/06/Wataya-Shaw-Cities.pdf

Wataya E, Shaw R (2022) Soft assets consideration in smart and resilient city development. Defining and debating on smart cities or smart buildings. Smart Cities 5(1):108–130. https://www.mdpi.com/2624-6511/5/1/7. https://doi.org/10.3390/smartcities5010007

Wataya E (2023) Integrating soft assets in smart and resilient city development in India. Unpublished PhD thesis, Keio University, Japan

Young A (1998) Measuring intangible investment, towards an interim statistical framework: selecting the core components of intangible investment. OECD Secretariat.

Chapter 4
The India Smart City Mission Program

4.1 Overview of the Smart City Mission Program

The India Smart City Mission program is a nationwide scheme designed to create urban sustainability launched by the Government of India in 2015 and supervised by the Ministry of Housing and Urban Affairs (MoHUA). Under the program 100 cities across India were selected to highlight their smart city development plans. The Government of India's objective was to address long-term challenges: providing basic infrastructure services, mitigating climate risks, and improving city resiliency against disasters.

The goal of India's smart city mission program is 'to promote cities that provide core infrastructure (Table 4.1), a clean and sustainable environment and provide quality of life to their citizens through the application of 'smart solutions' (MoHUA, Government of India, About Smart Cities mission). The program has the following areas of focus: (i) creating a smart city by building basic infrastructure to support economic growth, including physical, institutional, social, economic infrastructure; (ii) managing rapid urbanization; (iii) improving quality of life by providing effective access to health, education and promoting economic prosperity; (iv) application of smart solutions for electrical supply, traffic congestion by severance; efficient water supply and solid waste management; and (v) creating a clean and sustainable environment.

Although India believes that there is no standard definition of or template for smart city development, the Government adopted six fundamental pillars in the India Smart City Mission program (Fig. 4.1): (i) Community at the core; (ii) More from less (greater outcome with lesser/fewer resources); (iii) Corporative and competitive federalism; (iv) Integration, Innovation, Sustainability; (v) Technology as a means, not an end; and (iv) Sectoral and financial convergence.

E. Wataya and R. Shaw, *Soft Assets in Smart and Resilient City Development*, Disaster Risk Reduction, https://doi.org/10.1007/978-981-97-3373-6_4

Table 4.1 The core infrastructure elements

The core infrastructure elements (= BASIC INFRASTRUCTURE)	Some of the core infrastructure elements: adequate water supply, assured electricity supply, sanitation, including solid waste management, efficient urban mobility and public transport, affordable housing, especially for the poor, robust IT connectivity and digitalization, good governance, especially e-Governance and citizen participation, sustainable environment, safety and security of citizens, particularly women, children and the elderly and health and education

Source MoHUA, Government of India, Smart City Mission

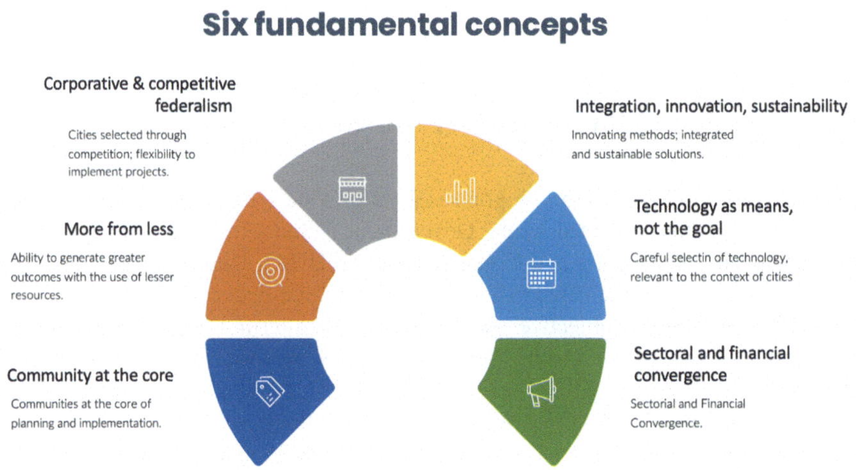

Fig. 4.1 Six fundamental concepts *Source* MoHUA, Government of India, About Smart Cities mission, https://smartcities.gov.in/about-the-mission

4.1.1 Key strategic areas

To achieve these broad objectives, the Smart City Mission's primary strategic purpose is to develop a comprehensive city plan consisting of Area Based Development (ABD) and Pan City Development—ABD focuses 'Redevelopment, Retrofitting, and Green field' development, and Pan City Development integrates smart solutions using a Command and Control Centre as the nerve center of city-wide smart solutions. Such comprehensive approaches endeavor to drive economic growth and improve the quality of life ('Strategy', Ministry of Housing and Urban Affairs, Government of India).

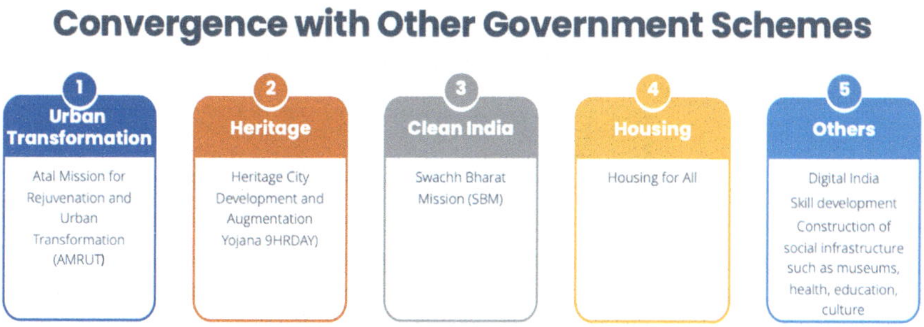

Fig. 4.2 Convergence with other government schemes. *Source* MoHUA, Government of India, Convergence

4.1.2 Key Elements of Implementation Scheme

Operation of the Smart City Mission program: The program is administered as a centrally sponsored scheme or CSS and the central Government provided financial support to the Mission of Rs. 48,000 crores over five years. Key funding sources for the Smart City Mission are the Central Bank, state governments, Urban Local Bodies or ULB, local public authorities, public–private partnerships and financial agreements with missions, programs, and schemes of the Indian Government, and other partners (MoHUA, Government of India, Financing).

Converging Smart City Mission with central or state government schemes: In order to promote smart city projects, and promote greater benefits derived from synergies between existing schemes and projects, the Government of India requires the merging and replication of the Smart City Mission project with other existing government schemes, e.g. AMRUT,[1] HRIDY, SBM, IPDS, Shelter for All, Digital India, make in India, Skill India (Fig. 4.2). Relevant external projects in human and financial resources and common activities were combined.

Monitoring: Each city selected in the program established a special purpose vehicle (SPV) to operationalize their Smart City Mission program. The SPV is led by an Apex Committee with a defined hierarchy of different groups (Fig. 4.3, and Table 4.2). Monitoring is undertaken at three levels: (i) national level, i.e. the Apex Committee (AC) and National Mission Directorates directing and monitoring national activities, fund approval, and release; (ii) the state level, i.e. High Powered Steering Committee to provide guidance and a platform for exchanging ideas; (iii) the city level, enabling collaboration among stakeholders. The major roles of the monitoring bodies are summarized in Table 4.2.

[1] Rejuvenation and Urban Transformation (AMRUT): Providing water tap connections to household in all statutory towns National Heritage City Development And Augmentation Yojana (HRIDY): preserving and revitalizing the rich cultural heritage; SBM: achieving universal sanitation coverage (open defection free, solid waste management improvement; IPDS: integrated power development scheme.

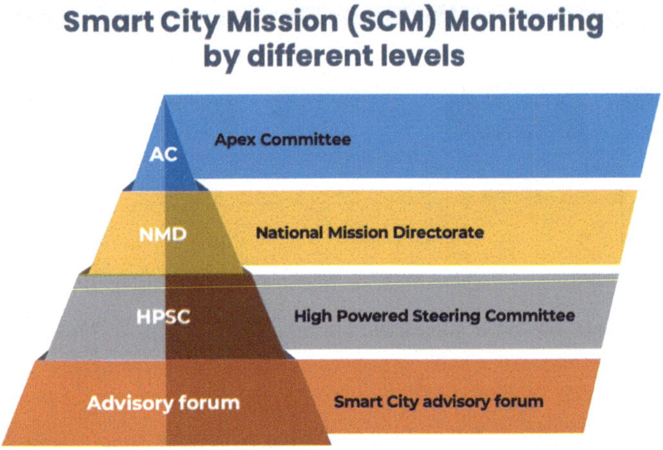

Fig. 4.3 Smart City Mission (SCM) monitoring by different levels . *Source* MoHUA, Government of India, Monitoring

Role of SPV: As a centralized implementation mechanism, SPVs have been established in each city. The SPVs are limited companies created by the states and municipal agencies to plan, appraise, approve, release funds, implement, manage, operate, monitor, and evaluate the Smart City development projects (Source: MoHUA, Government of India, SPVs).

The SPV is headed by a full-time chief executive officer, and its Board comprises nominees of the central, state, and local governments. The SPV has the flexibility to implement and manage the Smart City Mission projects and empowered to take any necessary measures. The SPV can also use consulting firms and appoint project management consultants (PMCs). The India Smart City website explains that the state and local governments (ULBs) shall provide: (i) a substantial revenue stream to ensure financial viability of the SPV, including sufficient creditworthiness to raise additional resources from the market, and (ii) a central government contribution to provide infrastructure resulting in public benefit outcomes.

On execution, the SPV manages projects as a joint venture (JV), a public–private partnership (PPP), utilizing a turnkey agreement formerly managed by the central Government. The unique feature of the SPV which uses a private CEO model is the ability to make quick decisions like the private sector. The SPV can also deal with communities and various stakeholders beyond the jurisdiction of city authorities.

A unique feature of the Government-supported Smart City Mission is leadership by an SPV which can cover operational and financial areas not undertaken by city authorities. However, there is also research on the operational capacity of SPVs. One of the findings is there is a shortage of adequate manpower or requisite legal jurisdiction to operate them. As a result, SPVs need to rely on state agencies to fill the operational gaps (Praharaj et al. 2017). Effective outcomes require strong coordination with adequate human capacity within the city bureaucracy.

Table 4.2 Smart City Mission (SCM) monitoring at different levels, roles

Apex Committee (AC)	Headed by the Secretary, MoUD • Approve the Proposals for SCM, monitoring their progress, release funds • Overall guidance and play an advisory role to the Mission
National Mission Directorate (NMD)	• Overall in-charge of all activities related to the Mission • Develop strategic blueprint and detailed implementation roadmap • Coordinate across Centre, States, ULBs and external stakeholders • Oversee Capacity building and assisting in handholding of SPVs, State, and ULBs
High Powered Steering Committee,	Chaired by the Chief Secretary • steer the Mission Programme in its entirety • Provide guidance to the Mission • Oversee the process of first stage intra-State competition • Review the SCPs and send to the MoUD for participation in the Challenge
Smart City advisory forum	• Advise and enable collaboration among various stakeholders • District Collector, MP, MLA, Mayor, CEO of SPV, local youths, technical experts, and at least one member from the area • Residents Welfare Association, Taxpayers Association/ Rate Payers Association, slum level federation, NGOs or • Mahila Mandali/Chamber of Commerce/Youth Associations • The Forum is called by CEO of SPV

Source MoHUA, Government of India, Monitoring

4.2 Disaster Risk Reduction

India is a severely disaster-prone country. It experienced about 430 major disasters and other hazards between 1991 and 2005 (UNDRR, Asian Disaster Preparedness Center 2020). In the World Risk Index (WRI) that measures the level of preparedness among 181 countries, India was ranked 89th in 2020 (jointly developed by the United Nations University Institute for Environment and Human Security (UNU-EHS), Bündnis Entwicklung Hilft (2022) and the University of Stuttgart in Germany). The World Risk Report (2022) also reported that India is the second highest risk country and seventh highest exposure to risks with complex interactions of multiple exposures and high intensities. This is also compounded by adverse climate change impacts.

Although smart city development aims to improve the quality of life of citizens and mitigate the impact of disasters, nevertheless the damage and destruction remains high according to the Disaster Risk Reduction Report. The Status Report 2020 (UNDRR, Asian Disaster Preparedness Center 2020), highlighted some key implementation issues which strengthen DRR capacity—lack of comprehensive and contextualized disaster information leading to warning risks of DRR and climate change adaptation into sectoral development; poor coordination among key agencies, especially implementing agencies, ministries, and other organizations; and limited capacity to prepare an implementation framework for risk reduction programs, activities and execution, funding, planning, and management. There is a clear need to strengthen disaster risk management planning and operations in the Smart City Mission program.

However, as UNDRR pointed out, 'climate resilience was not mentioned in a cross-section of India' Smart City Mission proposals (UNDRR, Prevention web, 2017). With increasing climate-induced disasters, it is important to address immediate disaster incidents critical to mitigate the risk of damage. It is also important to consider future impacts, not only focusing on taking preventive evaluation measures during actual disaster incidents. India launched a climate smart cities assessment framework 2.0 in 2020 (Table 2.2, Chap. 2), five years after the India Smart Mission was launched. It included more soft components such as communication aspects, awareness raising, behavioral change, comprehensive planning, and the use of a climate assessment reporting system. The addition of these soft assets elements would clearly complement the missing preventive and disaster mitigation elements of the India Smart City Mission. Closing these existing capacity gaps remains significant issue for the future development of Indian smart cities.

4.3 Selection of Two Indian Cities as Case Studies

The major reason for selecting the India Smart Mission program is its scalability to other Indian cities. Second, India is a disaster-prone country, with a large number of natural disasters involving different types of hazards across different climatic zones. The smart, resilient city plans provide potential solutions to these risks.

Two cities from the India Smart City Mission were selected as pilot studies for the application of the tool introduced in Chap. 3. The purpose was to drive economic growth and improve the quality of life of people by facilitating local area development and harnessing technology that lead to Smart Outcomes (MoHUA, Government of India, Smart City Mission). Another major reason for selecting the India Smart Mission program is its capacity to be adapted to other Indian cities. Bhubaneswar (located in the eastern coastal plains) and Nagpur (in the center of the Indian subcontinent) were selected as case study cities to represent diverse geographical locations in India.

4.3.1 *BHUBANESWAR*

4.3.1.1 City Profile

Bhubaneswar, the capital city of the State of Odisha, lies in a coastal region of the Indian Ocean. It is located in the eastern part of the State on the Kuakhai River. Bhubaneswar has many historical sites including ancient temples (it is known as the city of temples) (Fig. 4.4, Table 4.3). Ancient architecture spans 2,000 years from BC 3rd to AD fifteenth centuries ('Know Bhubaneswar, Bhubaneswar Smart City Limited'; Britannica). Bhubaneswar consists of two areas—old town and the modern quarter which developed from 1948 and, includes government buildings, universities, and other public places such as parks and gardens.

Fig. 4.4 Location map (Bhubaneswar). *Source* https://www.google.com/maps/place/Bhubaneswar,+Odisha,+India/@19.4950984,71.4062159,5z/data=!4m15!1m8!3m7!1s0x3a226aece9af3bfd:0x133625caa9cea81f!2sOdisha,+India!3b1!8m2!3d20.2375561!4d84.2700179!16zL20vMDFsNjln!3m5!1s0x3a1909d2d5170aa5:0xfc580e2b68b33fa8!8m2!3d20.2960587!4d85.8245398!16zL20vMDNjenFz?entry=ttu

Table 4.3 City profile (Bhubaneswar)

City	• Bhubaneswar (State of Odisha)
Area	• Metropolitan area: 186 sq. km (72 sq ml) • Metro are 1,110 sq. km (430 sq ml.)
Government bodies	• Bhubaneswar Development Authority (BDA)—urban planning, land development, infrastructure projects within the city of Bhubaneshwar • Bhubaneswar Municipal Corporation (BMC)—municipal services and local governance
Population (2011) ppl	• Metropolis: 974,000 • Metro: 1.3 Mill
Climate geology	• Tropical weather climatic conditions, Eastern coastal plains
Type of disasters	• Cyclone, floods, Tsunami • Hear wave

Source Bhubaneswar Municipal Corporation; Bhubaneswar smart City Limited; Bhubaneswar India Local Governance Network; Bhubaneshwar Development Authority

The city plays an important role as a regional gateway to the tourist areas of Puri, Konark, and Chilika Lake. Because of its coastal location, it is also a trading hub connecting other countries in Southeast Asia. In addition, the city also connects inland areas with the national highway to Kolkata in the north and Chennai, and is also linked to other major cities by subsidiary road networks.

4.3.1.2 Smart City Mission

Smart city vision and strategic focus

Bhubaneswar was selected as part of the Smart City Mission in 2016 (round 1) and established a special purpose vehicle (SPV) called Bhubaneswar Smart City Limited to implement the smart mission and its programs.

Urban development plan: Bhubaneswar's population grew from 15,612 in 1950 to 394,641 in 1990, then, 638,000 in 2000 and 1,226,033 in 2022 (World Population Review). In the 50's Bhubaneswar's target population as an administrative center was 40,000 due to the rapid population growth, and an interim development plan was prepared by the Bhubaneswar Development Authority (BDA) in 1993, but with economic growth from the 1990's onwards, the city expanded considerably. To address this growth, BDA prepared a comprehensive urban development plan 2030 for land use planning, setting out the spatial strategy for the 14 planning sub-areas identified in the plan to meet the different needs of each region. There are also other strategic plans for specific purposes, including a smart city mission proposal that was prepared by BDA and BMC with support from international consulting firms. However, Bhubaneswar has been facing issues such as unplanned urbanization in the peri-urban regions, a lack of essential infrastructure, unauthorized construction activities leading to increased flood risks, groundwater depletion, and a reduction of green spaces. Consequently, as the population continued to grow, the city continued to face challenges related to water supply, sewerage management, solid waste collection, and stormwater drainage (Chatterji 2023).

Smart City Mission: The launch of the Smart City Mission in 2015 presented many Indian cities with a new opportunity to design forward-looking, transformative plans. Bhubaneswar Development Authority (BDA) and Bhubaneswar Municipal Corporation (BMC) developed the Smart City Plan, with strategic focus on the following: (i) creating a sustainable model based on new urban principles that provide citizens with prosperity, safety, and equity; and (ii) expansion of the urban planning system to cover integrated land use and transport planning, infrastructure planning, and socio-economic planning. Underpinning these focus areas, the plan embraces five strategic pillars—responsive governance, transit-oriented development, financial sustainability, infrastructure, and socio-economic development (Bhubaneswar Smart City Mission Proposal, Stage 2).

According to the proposal, Bhubaneswar has additional major urban development challenges, including: (i) urban sprawl and leapfrog development due to low density development and limited release of public lands for development in its core urban area; (ii) with a lack of a city core, no business district has evolved over the years in the city nor been able to attract private sector investment; (iii) a weak public transport system with the city bus service network covering limited areas of the city and a smaller number of busses compared with national average; (iv) 36% of city population live in unregulated informal settlements due to a shortage of affordable housing options; (v) encroachment of unhygienic natural drainage zones; (vi) uncontrolled tourist traffic; and (vii) a multiplicity of agencies (Source: Bhubaneswar Smart City Mission Proposal, Stage 2).

With this background, the Bhubaneswar city vision is as follows:

- Transit oriented city with compact urban center that promotes active, connected, and sustainable mobility choices
- Livable city environment providing a diverse range of housing, educational, and recreational opportunities while enhancing its heritage, arts, and traditional communities
- Child-friendly city providing accessible, safe, inclusive, and vibrant public places
- Eco-city co-existing in harmony with natural resources to foster a resilient, clean, green, and healthy environment
- Regional economic center attracting knowledge-based enterprises and sustainable tourism activities by leveraging and empowering its institutions, local businesses, and informal workforce
- The above actions are implemented through responsive governance, core values of equity, participatory decision-making, transparency through accessible information and technology

Based on the study by Chatterji (2023), Bhubaneswar's smart city proposal encompasses not only smart solutions for city-wide infrastructure, urban renewals, and retrofits, but also integrates social or soft components to enhance equity, accessibility, and economic opportunities for its citizens. This approach supports Bhubaneswar's

Table 4.4 Bhubaneswar smart city key features

Bhubaneswar's vision	To be a regional economic center, livable, ecologically sustainable, efficient transport, child-friendly city with accessibility and safety, through participatory decision-making, responsible governance and open access to information and technology
Major interventions (ABD, Pan City Development)	• The town center district (985 acres) development for ABD (A walkable, well-connected mixed-use area; A model of low-impact carbon neutral development) • Social equality and capacity development in ABD development area • Intelligent City Operations and City Management Center (ICOMC) development for Pan City
Smart city implementation body	Bhubaneswar Smart City Ltd (BSCL)
Project financing approaches	Central government, Government of Odisha, Bhubaneswar city, Converge scheme (AMRUT、Swacch Bharat), PPP, and others

Source The Smart City Challenge Stage 2 Bhubaneswar 2016

extensive consultation process involving five steps- Inform, Consult, Involve, Collaborate, Empower- that were undertaken in 'options identification'; 'prioritization'; and 'planning' stages (Bhubaneswar Smart City Mission Proposal, Stage 2). This helps the city receive feedback during the implementation stage (Table 4.4).

4.3.2 NAGPUR

4.3.2.1 City Profile

Nagpur is a city in the State of Maharashtra, located in western India. The city lies along the Nag River and is situated in the middle of the Indian continent. It is a fast growing metropolis, the third largest city in Maharashtra, and a major commercial and political center of the Vidarbha region of Maharashtra. The city is also known as 'Orange City' because it is a major orange growing region (https://nagpur.gov.in/history/) (Fig. 4.5, Table 4.5).

The city is managed by the Nagpur Municipal Corporation (NMC) led by a Mayor supported by several committees, including the standing committees on health and sanitation, education, water works, public works. The Nagpur Metropolitan Regional Development Authority (NMRDA) supervises metropolitan development comprising 721 villages across nine local administrative units or tehsils in the district. The city is also an educational and cultural center for the region.

Fig. 4.5 Location map (Nagpur). *Source* https://www.google.com/maps/place/Maharashtra,+India/@19.5739444,70.5091692,5.11z/data=!4m6!3m5!1s0x3bcfc41e9c9cd6f9:0x1b2f22924be04fb6!8m2!3d19.7514798!4d75.7138884!16zL20vMDU1dnI?entry=ttu)

On transport, Nagpur has effective connectivity with all the important adjacent domestic cities. All major highways (national highways 6, 7) and railway routes to Mumbai, Chennai, Delhi pass through the city. An electrified broad gauge commuter railway track connects Nagpur to the four major metropolitan areas of India. Its international airport is one of the busiest in the country.

4.3.2.2 Smart City Mission

Smart city vision and strategic focus

Nagpur was selected as part of the Smart City Mission in 2016 (round 2) and NMC established an SPV called the Nagpur Smart and Sustainable Development Corporation Limited (NSSCDCL) to implement the mission goals and programs.

Table 4.5 City profile (Nagpur)

City	• Nagpur (State of Maharashtra)
Area	• Nagpur city area: 217.65 sq km • Nagpur district area (9,897 sq km)
Government body	• Nagpur Municipal Corporation (NMC)—local government bodies in the city, municipal administration and governance • Nagpur Metropolitan Regional Development Authority (NMRDA)—planning and development activities in the Nagpur Metropolitan Region, development and regional planning
Population (2011) ppl	• City (Metro?) area: 2,405,421 • Nagpur district: 4.653,570
Climate geology	• Dry
Type of disasters	• Flood, Landslide, drought, heat wave

Source Nagpur Municipal Corporation; District Nagpur, Government of Maharashtra; District Disaster Management Authority Nagpur; Nagpur Metropolitan Regional Development Authority

Regional plan: The Nagpur Regional Plan for 1970 to 1990 (910,010 in 1990) addressed a significant industrial expansion, providing employment of around 40,000. The Maharashtra Industrial Development Corporation (MIDC) acquired 750 hectares to (i) upgrade the industrial area; and (ii) develop roads and water infrastructure. However, the development in this period revealed the problems of limited city development requiring that the planning area be expanded to a larger area (administrative district). Issues that developed in the Nagpur city and surrounding regions were housing shortages and increasing slums due to growing population (472,859 in 1950 to 2,991,427 in 2022) (World Population Review), unauthorized construction, and traffic failures. Consequently the plan involved dividing the entire region into a structured hierarchy of settlements to facilitate decentralization and shifting the focus of development and population away from the Nagpur urban cluster towards potential satellite growth areas (Nagpur Improvement Trust 2015).

The NMC prepared the City Development Plan for Nagpur, 2041 in 2015. A total of 12 sectors were incorporated into the Plan focusing on urban roads, traffic and transportation, water, sewage, and sanitation, storm water drains, solid waste management, slum housing, heritage development, tourism, urban governance, social infrastructure, urban environment, urban governance, social infrastructure, and disaster management (Ministry of Urban Development, Government of India 2015).

Smart City Mission: The Smart City Mission offers Indian cities opportunities to enhance sustainability, livability, and innovation in line with their urban visions. The program becomes an integral part of their existing plans, driving progress towards mission goals. According to Nagpur's Smart City Mission proposal, Nagpur's major city development challenges are (i) multiple and duplicative planning institutions and dichotomy in growth causes uncoordinated and ineffective planning affecting quality of life; (ii) unplanned and haphazard development due to inconsistent development planning and land use planning; (iii) inequitable distribution of civic services due to unplanned layout and inadequate water supply, sanitation, sewerage, public transport and street lighting, and others; (iv) inadequate public transport system, including bus services which is the only available city public transport; (v) administrative and operational inefficiencies which lower the quality of public service delivery; and (vi) lack of employment opportunities due to the slow pace of economic growth with educated people unable to find adequate employment opportunities (India Smart City Mission. The Smart City Challenge Stage 2—Nagpur, 2018).

With this background, Nagpur's city vision is as follows:

Nagpur seeks to transform the city into the most livable 'eco-friendly edu-city', that connects citizens virtually with the government to create an inclusive ecosystem loop. The city has adopted a transformative agenda with four key features in its action plan to enhance quality of life in Nagpur, i.e. smart living, smart mobility, smart environment, and smart governance. It is supported by 12 areas of focus for action: inclusive living, policy-centric city, safe and walkable streets, economic vitality, TOD, connect places and effectively move people, carbon neutral and sustainable habitat, Swachh Nagpur, urban greening, digital efficiency and good governance, revive Nagpur's image as regional economic center, and boost job creation (India Smart City Mission. The Smart City Challenge Stage 2—Nagpur, 2018). Nagpur's smart city proposal places strong emphasis on ITC and technology integration to enhance the efficiency of city services and smart infrastructure to create sustainable economic growth (Tables 4.6 and 4.7).

Table 4.6 City vision and strategic blueprint area

1. Smart Living	2. Smart Mobility	3. Samrat Environment	4. Smart Governance
Inclusive living	Transit Oriented Development	Carbon Neutral & Sustainable Habitat	Digital Efficient and Transparent Governance
Poly-centric city	Connect Places and Move People	Swachh Nagpur	Revive Nagpur's image as the regional economic center and boost job creation
Safe and walkable streets	Urban Regeneration with walk-to-work	Urban Greens	
Economic vitality			

Source Nagpur Smart City Mission Proposal, Stage 2—Nagpur, 2018

Table 4.7 Nagpur smart city key areas

Nagpur's vision	'To transform India's heart-Nagpur into the most livable eco-friendly, edu-city that electronically connects people with the government to co-create an inclusive ecosystem'
major interventions (ABD, Pan City Development)	• Retrofitting of Pardi-Bharatwada-Punapur in the eastern periphery (951 acres) to make the city inclusive growth and enhance the city's profile by improving the quality of life in vulnerable areas • Invest in a customized ICT-based Smart Swachh City Solution to streamline city's garbage management based on the vision for a 'Clean' city • 'Safe City' by the Nagpur City Community Network (NCCN) and & a Unified Operations Command and Control Center (UOCCC) to operationalize ICT-based solutions • Satellite imagery-based monitoring of city expansion to check unplanned development
Smart city implementation body	Nagpur Smart and Sustainable Development Corporation Limited (NSSCDCL)

Source Nagpur Smart City Mission Proposal, Stage 2—Nagpur, 2018

4.4 The Ease of Living Index and the Municipal Performance Index

The Government of India published two performance indicators in 2020. One is the Ease of Living (EOL) Index and the other is the Municipal Performance Index (MPI). The EOL provides performance measurement of citizen well-being and the MPI, the effectiveness and quality of urban governance.

The EOL index was launched in 2018 to assess the various Indian flagship programs in operation, i.e. Smart Cities Mission, Swachh Bharat Mission (SBM), Atal Mission for Rejuvenation and Urban Transformation (AMRUT), Pradhan Mantri Awas Yojana-Urban (PMAY-U), Deen Dayal Antyodaya Yojana-National Urban Livelihood Mission (Day-Nulm), and Heritage City Development and Augmentation Yojana (HRIDAY). Most of them are listed as convergence programs in the national-wide Smart City Missions. EOL provides evidence-based planning and assessment of strategies and programs implemented to produce a better quality of life for its citizens. The MPI was developed in the EOL index reform process. A clear distinction was drawn in the EOL emphasizing the outcomes of development, with key three pillars: quality of life, economic ability, and sustainability. In addition, there is a further sub-division into 14 categories and a total of 50 indicators. Also the EOL is closely linked to the UN Sustainable Development Goals (SDGs) (MoHUA 2021c; Ease of Living Index 2020, Assessment Report, https://livabilitystore175634-prod.s3.amazonaws.com/public/docs/Ease_of_Living_Report.pdf). The MPI represents the inputs or enabling factors separately assessed under MPI

(MoHUA 2021d; MPI 2020, https://livabilitystore175634-prod.s3.amazonaws.com/public/docs/MPI_Report.pdf). The assessment tool identifies the gaps in policies and implementation reflected in the outcomes. There are five key pillars: service, finance, planning, technology, and governance and it sub-divides 20 sectors and into a total of 100 indicators.

Table 4.8 compares each index that supports the Smart City Mission by providing measurements of service quality and enhancement of citizens' well-being. Each index assesses how soft assets are considered and evaluated in the planning stage and utilized in the implementation stage. The EOL index addresses outcomes. MPI's assessment of the performance of urban governance is also important to measure project effectiveness. In this chapter, the selected EOL indicators and methodologies will be examined and used to analyze the relationship between soft asset consideration and performance.

4.5 Ease of Living Index Indicator Overview

4.5.1 Framework

The Ease of Living Index has four pillars with different levels with indicators for each shown in Table 4.8 in the previous section. The percentage allocation is 35% for Quality of Life, 15% for Economic ability, 20% for Sustainability, and 30% for Citizen's perception survey. According to EOL 2020, each pillar set out in Table 4.9 covers the following areas:

4.5.2 Methodology

The examination used several approaches to capture optimal results taking into consideration diverse development stages of different municipalities. Two groups of cities were classified by population greater than 1 million and less than 1 million. 111 cities were reviewed for selection and include all cities covered by the Smart City Mission.

Different scoring methods were applied to each indicator. At the stage of Category Scores, each indicator under the category has been given the same weight. The weights for the separate pillars were assigned based on inputs from experts and set proportionally across the pillars. The category values are calculated by adding up the weighted scores with specific formulae for the Ease of Living Index. Capacity pillars and key index sectors are shown below (Figs. 4.6, 4.7, 4.8, 4.9).

Table 4.8 Comparison between Ease of Living (EOL) Index and Municipal Performance (MPI) Index

	Formulated by	Cities	> 1 mil	< 1 mil	basis of data	4 broad city capacity pillars (Indicators)	Characteristics selected indicate
Ease of Living Index (EOL) 2020 https://livabilitystore175634-prod.s3.amazonaws.com/public/docs/Ease_of_Living_Report.pdf	MoHUA with the support of their knowledge partner—Institute of Competitiveness. Other partners in this venture are Smart Cities Mission Directorate, Karvy Data Management Services, National Informatics Centre,	111	49	62	50 indicators	- Quality of life (education, health, housing and shelter, water, sanitation, hygiene, solid waste management, mobility, safety and security, and recreation); - Economic ability (level of development, economic opportunities); - Sustainability (environment, green space and buildings, energy consumption, and resilience) - Citizen perception survey (conducted between February and March 2020 to confirm results of data)	Helps in knowing **about the well-being of the citizens** It informs city governments about the gaps in urban policies, planning, and implementation initiatives, and offers an opportunity to improve the lives of citizens, as well as track the status of the sustainable development goals (SDGs)
Municipal Performance Index (MPI) 2020 https://livabilitystore175634-prod.s3.amazonaws.com/public/docs/MPI_Report.pdf	City Data Officers, Smart City Consultants, and Smart City Fellows	111	51	60	20 sectors, 100 indicators	- Governance (transparency and accountability, - Human resources, participation, and effectiveness); - Technology (digital governance, access, and literacy); - Planning (plan preparation, implementation, and enforcement); - Services (education, health, water and wastewater, solid waste management and sanitation, registration and permits, and infrastructure); and Finance (revenue and expenditure management, fiscal responsibility, and decentralization)	**Describe the quality of urban governance** (or performance of municipalities) in select cities The data and analysis provided in the report will help municipalities in knowing the planning and management practices that require strengthening/reform

Source MoHUA, Government of India, 2021c, 2021d, https://livabilitystore175634-prod.s3.amazonaws.com/public/docs/Ease_of_Living_Report.pdf https://livabilitystore175634-prod.s3.amazonaws.com/public/docs/MPI_Report.pdf

Table 4.9 Index framework and weight

Pillars	Quality of Life (35%)	Economic ability (15%)	Sustainability (20%)	Citizen's perception survey (30%)
Capture/ evaluate	Understand the different aspects contributing to a decent urban life The goal is to assess a holistic impression of the quality of life in India's urban cities	Economic well-being of citizens by evaluation the level of economic development and inequalities that they encounter in a particular city	Along the lines of availability of green spaces, promotion of green buildings level of energy consumption, the quality of natural resources such as air and water and the city's ability to withstand natural disasters	Perception of the city residents and allows them to evaluate the level and quality of development
Approach	Examining provisions for necessities (urban services)		Data provided by the cities and validated by secondary sources	Survey Results were used for validation of data provided by cities

Source MoHUA, Government of India, 2021c Ease of Living Index 2020

Fig. 4.6 Ease of Living Index 2020: Bhubaneswar. *Source* MoHUA, 2021c Ease of Living Index (EOL), https://livabilitystore175634-prod.s3.amazonaws.com/public/docs/Ease_of_Living_Report.pdf

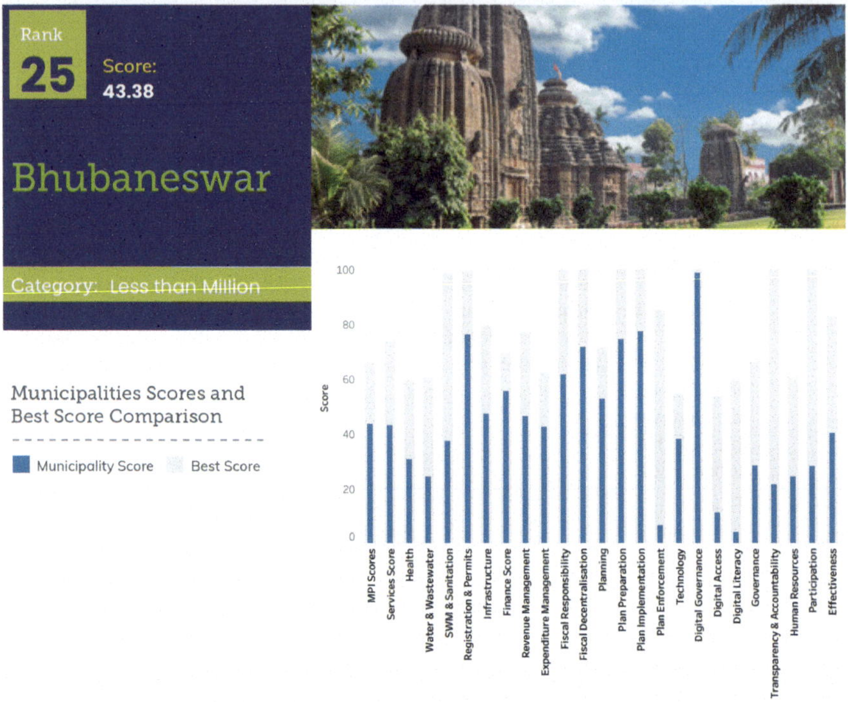

Fig. 4.7 Municipal Performance Index 2020: Bhubaneswar. *Source* MoHUA, 2021d Municipal Performance Index (MPI), https://livabilitystore175634-prod.s3.amazonaws.com/public/docs/MPI_Report.pdf

4.5.3 Ease of Living—Citizen's Perception

The citizen's perception of their environment was an important component of the Ease of Living Index used to directly capture the citizen's perspective about the quality of city life. As shown in Table 4.9, the index framework and weight in Sect. 4.5.1., the Citizen Perception Survey is a very useful indicator to validate the data provided the city authority uses it as a tool to identify gaps from the citizen's experiences and assessments of city service delivery and the effectiveness of the prevailing public administration. The survey called '#MyCityMyPride' was conducted by MoHUA, the National Institution of Urban Affairs and other partners (The National Institution of Urban Affairs, EASE of LIVING 2020, Citizen Perception Survey). The extensive consultation process helped Bhubaneswar conduct a comprehensive survey.

Fig. 4.8 Ease of Living Index 2020: Nagpur. *Source* MoHUA, 2021c Ease of Living Index (EOL), https://livabilitystore175634-prod.s3.amazonaws.com/public/docs/Ease_of_Living_Report.pdf

4.5.3.1 Methods and Approaches

The survey method combined online and offline inputs: from January to February 2020 online and to March by in-person survey (Citizen Perception Survey in The Ease of Living 2020). Various modes of communication (newspaper, social media, billboard, outdoor advertising, campaign) were introduced to promote the survey among citizens, including 'MyGov.in.' a citizen engagement platform funded by the Government of India that involves citizen participation in assessing the performance of city governance and development. Also some mechanisms were introduced by the Government to incentivize the level of participation.

4.5.3.2 Questionnaires

The survey questions were simple and focused on 'service provision' to evaluate how citizens perceive their satisfaction with service delivery: (i) Quality of lives: housing, cleanness, water supply, commuting, mobility, safe for women, culture, amusement facilities; (ii) Economic ability: Job opportunities, cost of living, accessibility to financial services; and (iii) Sustainability: air quality, green places, and green electricity supplies.

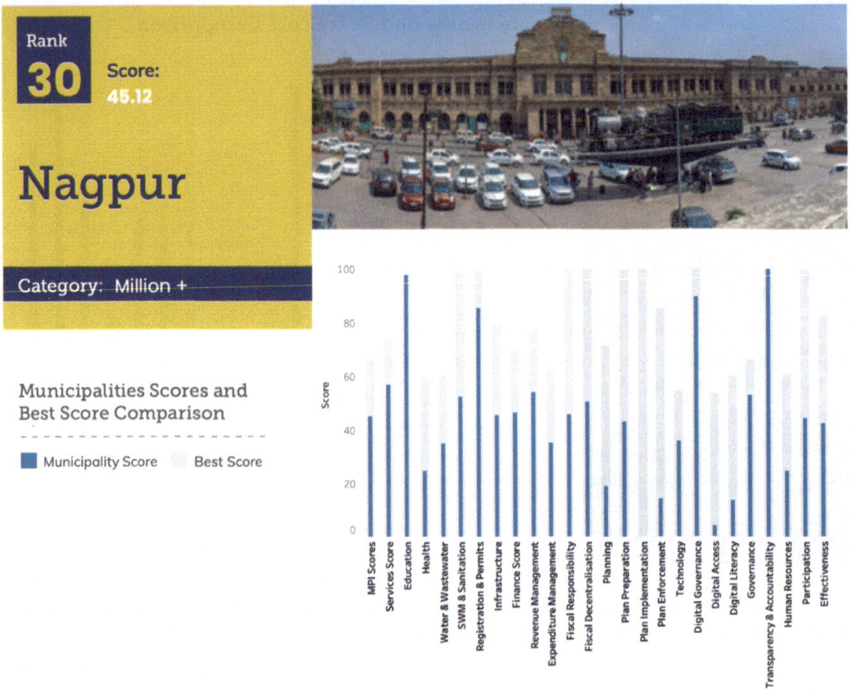

Fig. 4.9 Municipal Performance Index 2020: Nagpur. *Source* MoHUA, 2021d Municipal Performance Index (MPI), https://livabilitystore175634-prod.s3.amazonaws.com/public/docs/MPI_Report.pdf

4.5.3.3 Number of Respondents and Groups

On the online portion both 'MyGov.in.' and on the other online part, more than 2.5 million people participated. The in-person or face-to-face citizen interviews were conducted by experienced field enumerators and involved more than 95,000 samples in total. Hence the government received 2.65million responses. In terms of the demography of respondents, males between age 18–35 were the largest group (41.4%) followed by males over age 35 groups (31.9%), females between age 18–35 group (15.7%), then females over age 35 group (11%). Table 4.10 and Table 4.11 show the results of the indexes of the two case study cities selected for this book.

Among all cities (111 cities), Bhubaneswar is ranked 1 (CPS score: 94.8; EOL Index without CPS: 31.42; Nagpur is ranked 40 (CPS score: 78.1; EOL Index without CPS: 31.90) (MoHUA, Government of India, Ease of Living Index 20,201). Under the 'Quality of Life', category, the ranking was divided into two groups based on population size, i.e. more than million and less than million. As for Bhubaneswar, it is ranked 21 out of 69 cities (score: 51.79). Nagpur is ranked 38 out of 49 cities (score: 50.59). Among other index scores of the LOE, the CPS is relatively higher

Table 4.10 Ranking of Ease of Living Index 2020

Population category	National average	Bhubaneswar		Nagpur	
		Less than million	Score	Million +	Score
Ranking: **Ease of Living Index (EOL) 2020**					
Ease of Living Index		2	59.85	25	55.33
QUALITY OF LIFE	51.38	21	51.79	38	50.59
ECONOMIC ABILITY	13.17	19	11.57	20	15.35
SUSTAINABILITY	53.65	11	57.77	16	59.43
Ease of Living Index (w/o CPS)*	30.69		31.42		31.90
CITIZEN PERCEPTION SURVEY*	76.08	1	94.80	40	78.10
Ease of Living Index	53.51	2	59.85	25	55.33

* Ranking under all cities

Table 4.11 Ranking of Municipal Performance Index 2020

Verticals	National average	Bhubaneswar		Nagpur	
		Less than million	Score	Million +	Score
Ranking: **Municipal Performance Index (MPI) 2020**					
National average	43.13				
GOVERNANCE	42.83	52	28.62	15	52.94
PLANNING	34.03	3	52.76	50	19.16
TECHNOLOGY	24.02	3	37.92	9	35.75
FINANCE	51.11	17	55.52	43	46.27
SERVICE	52.13	45	43.18	22	56.80

Source MoHUA, 2021c, 2021d Government of India, Ease of Living 2020. Municipal Performance Index 2020

than other cities. At least both are in the first 35% of total cities, whereas the quality of life scores is relatively lower than the CPS.

Each city made efforts to increase the response rate of the Citizen Perception Survey. For example, Bhubaneswar Smart City Limited, in collaboration with Bhubaneswar Municipal Corporation (BMC) and BDA, carried out an extensive publicity campaign through several communication channels to motivate citizens to participate in the survey (Odisha Bytes 2021). Effective and broad-based campaigns can motivate its citizens to participate more resulting in higher rankings. The results reflect the citizen's perception with a relatively high degree of accuracy based on the results of other pillars such as Quality of Live, Sustainability, which resulted in higher 'Ease of Living Index' results.

Interestingly, each city's top five City Scores are similar—Bhubaneswar has 'City resilience'; 'Housing and shelter'; 'Safety and security'; 'Energy consumption'; 'Education'; and 'Security', while the lowest item was 'Level of economic development'. Nagpur has 'Safety and Security'; 'City resilience'; 'Education'; 'Housing and shelter'; and 'Energy consumption' in descending order according to the City Scores, while the lowest item was 'Recreation'. The next chapter examines the modes and types of soft assets identified.

References

Bündnis Entwicklung Hilft (2022) The World Risk Report, Ruhr University Bochum—Institute for International Law of Peace and Armed Conflict (IFHV), https://reliefweb.int/report/world/worldriskreport-2022-focus-digitalization

Chatterji T (2023) Planning and development of Smart Cities: a case study of Bhubaneswar, Institute of Town Planners, India Journal 20 x 2, April—June 2023. https://www.researchgate.net/publication/372890730_Planning_and_Development_of_Smart_Cities_A_case_study_of_Bhubaneswar

Ministry of Urban Development, Government of India (2015) City development plan for Nagpur city—2041 final report, March 2015. https://www.metrorailnagpur.com/pdf/Final%20CDP_Nagpur%20-Mar%2015%20of%20NMC.pdf

Ministry of Housing and Urban Affairs, Government of India, (2021c), Ease of Living Index 2020: Assessment Report https://livabilitystore175634-prod.s3.amazonaws.com/public/docs/Ease_of_Living_Report.pdf

Ministry of Housing and Urban Affairs, Government of India, (2021d) Municipal Performance Index 2020: Assessment report, https://livabilitystore175634-prod.s3.amazonaws.com/public/docs/MPI_Report.pdf

Nagpur Improvement Trust (2015) Draft development plan report, Published u/s 26 of the Maharashtra Regional and Town Planning Act, 1966. http://www.nmrda.org/pdf/Metro_Region_DCR.pdf

Odisha Bytes (2021) Ease of Living Report: Smart city Bhubaneswar Tops in Citizen Perception Survey, June 19. https://odishabytes.com/ease-of-living-report-smart-city-bhubaneswar-tops-in-citizen-perception-survey/

Praharaj S, Han H, Hawken S (2017) Urban innovation through policy integration: Critical perspectives from 100 smart cities mission in India. Faculty of Built Environment, University of New South Wales, Sydney. https://www.researchgate.net/publication/317986481_Urban_innovation_through_policy_integration_Critical_perspectives_from_100_smart_cities_mission_in_India

The United Nations Office for Disaster Risk Reduction, Asian Disaster Preparedness Center (2020) Disaster risk reduction in India 2020. Status Report. https://reliefweb.int/report/india/disaster-risk-reduction-india-status-report-2020, UNDRR

Chapter 5
Utilization of Soft Assets: Case Studies

5.1 Applying a Qualitative Framework Tool to Identify and Measure the Use of Soft Assets

This chapter reviews how soft assets are incorporated in both the planning and implementation stages of the Smart City Mission program proposal, including:

(i) Identifying where soft assets are incorporated;
(ii) The findings about the characteristics and integration of soft assets from two city cases; and
(iii) Explaining the similarities and differences of the approaches between the two cities.

As discussed in Chap. 3, due to the unique characteristics of soft assets and the difficulty of establishing clear linear relations between inputs and outputs, it is too complicated to trace the origins of many embedded soft assets used by cities. In addition, unlike the private sector, the ultimate goal of a city is to enhance the well-being and quality of life of its residents through the efficient delivery of essential services, effective governance and alignment with the city's mission, vision and objectives. In this context, quantifying results in monetary terms proves to be challenging and identifying and assessing the value of soft assets is complicated.

Soft assets drive value creation in an urban environment. Understanding this process, including its underlying mechanisms, awareness, insight, and approach to using soft assets, is key to improving the service performance of the city and better managing the ultimate outputs of soft assets. The holistic aspect of how soft assets are integrated are discussed in this chapter. The findings can be positively analyzed using the Ease of Living Index 2020 discussed in Chap. 4. These assessments are also related to the analysis of the research findings in Chap. 6 supporting the basic assumption set out in Chap. 1 Figs. 1.2 and 1.3 that incorporating soft assets in smart and resilient cities create value.

E. Wataya and R. Shaw, *Soft Assets in Smart and Resilient City Development*, Disaster Risk Reduction, https://doi.org/10.1007/978-981-97-3373-6_5

5.1.1 Survey Approaches

The survey of respondents from the two Indian smart city case studies examined how city authorities incorporated soft assets into smart, resilient city project planning and implementation.

Desk top research methods (Proposal stage)

Objective

The desk top research examined how soft assets consideration and decision-making works in actual cases on the ground. A desktop review of part (a) to (c) of the India Smart Mission proposal (Stage II) was undertaken. The proposal comprised: (a) the City Profile; (b) the Area Based Proposal (redevelopment, retrofitting, and green field); (c) the Pan-City Proposal (integration of smart solutions); (d) the Implementation Plan; and (e) the Financial Plan. Note that the City Profile section includes both past planning, current status and future planning. *The qualitative framework to identify an adaptation of soft asset incorporation* (Table 3.8, Chap. 3) was applied to examine the Smart City Mission proposal. The data collected was then collated and organized using an analytical tool, the Cytoscape (https://cytoscape.org), to identify nodes that produced influencing factors to create value over time.

Using the network analysis tool, the degree of soft asset consideration was examined from two angles: First is to identify the relationships between core elements ("large node") from the key nine elements namely: (i) resource management; (ii) mobility (accessibility); (iii) health, well-being; (iv) equal access to services; (v) smarter services via technology; (vi) community engagement; (vii) quality communications; (viii) security; (ix) resilience (Fig. 3.6, Chap. 3), and their connectivity structures with actions/initiatives ("small node") linked (in the 'Cystoscope', it is 'relation' or 'edge') by soft asset capital components namely: (i) people/human; (ii) institutional; (iii) external; (iv) technology, information; (v) social capitals (Table 3. 5, Chap. 3). The resulting data was analyzed and categorized, including which soft asset elements and capital components were incorporated into each of the 45 items above.

These results demonstrated which areas the city emphasized more than others, which elements were more influential over the entire connectivity, and also the potential synergetic effects interacting to create additional value in the future. Second is identifying the level of consideration of soft asset capitals and observation of the degree of alignment with and potential gaps in the city's vision and objectives. In addition, surveys were conducted to: (i) understand city administrators and key stakeholders' experiences, observations, insights; and (ii) examine whether the survey results validate the desk top research and analysis of the proposal stage.

Survey methods

Limits of the Survey

Due to the intrinsic nature of soft assets and the broader categorization of indicators that are used in *the qualitative framew*ork, the review and identification of soft asset consideration in the Smart City Proposal involve degree of subjectivity. To identify initiatives or activities in the proposal, a key word search function was also used comprising 45 items of indicators (Table 3.8, Chap. 3). The author also applied five soft assets capital components to initiatives or activities under the nine key value elements from several different angles.

In-person interview and survey questionnaires (implementation stage)

This research also used in-person interviews and survey questionnaires targeting influential decision makers and stakeholders in the two city studies—Bhubaneswar and Nagpur. As some topics were difficult, additional explanations were necessary tailored to each individual's level of understanding. Also, the survey was conducted by a native speaking interviewer with whom the author of this research closely communicated to customize appropriate interview questions for each city respondent.

Objectives

The objectives of the survey were to understand: (i) how a senior city official, implementer, academia, NGO staff, and private sector employees made decisions about using soft assets during the planning and implementation stage of the Smart City Mission; (ii) the observations and insights of the less senior decision makers; (iii) how a balance between soft and hard assets is achieved within the decision-making process; and (iii) how the soft assets created value and impact on the functioning of the Smart City Mission.

Survey methods

Both structured and unstructured interviews were used to obtain detailed responses from target respondents. The survey form targeted key respondents within a pre-set time frame.

- In Bhubaneswar, the city public relations officer initially met the interviewer to discuss the questions then collected inputs and feedback from appropriate municipal departments to ensure more focused responses.
- In Nagpur, the respondents reviewed the questionnaires prior to the meetings. The local interviewer ensured that the respondents fully understood the context and nature of the questions.

Target groups

The following target groups were selected and interviews were conducted for 5 days in each city (Table 5.1).

Table 5.1 Affiliations of the interviewees

City	Target interviewee/responded	Type of affiliation
Bhubaneswar	Director	Technology Consultant to BSCL
	Youth Specialist cum Programme Coordinator	BSCL (Supported by UNPFA)
	Urban Planner and Social Media Manger (Respectively)	Bhubaneswar Urban Knowledge Centre (BUKC)
	Technical Advisor- Climate Smart Cities	GIZ
	Program Manager	Private sector
	Associate- General Manager	Private sector
	Assistant Professor	Academia
	Assistant Professor	Academia
Nagpur	Project executive	Nagpur Smart City CDCL (SPV)
	Technical officer (Engineer)	Nagpur Smart City CDCL (SPV)
	Additional Commissioner (General)	Nagpur Municipal Corporation (NMC)
	Assistant manager (energy & climate)	Non-Profit Organization
	Manager	Private sector
	Director	Civil society
	Regional director	Civil society

Source Wataya (2023)

5.1.2 Survey Questionnaires and Limitations of the Interview Survey

The survey questionnaire consisted of open-ended questions covering the Bhubaneswar Smart City Mission and the Nagpur Smart City Mission programs. The survey conducted in each city endeavored to capture an overall picture of how city officials perceived soft assets and how they evaluated their utility in the Smart City Mission implementation stage, including performance issues and creating value. Tracing every activity identified at the proposal stage analysis was not feasible. It was also difficult to trace and identify actual occurrences of synergetic, spillover, and scaling up effects. However, it was possible to compare the proposal stage and implementation stage to evaluate and gain a picture of the performance of soft assets and how values are created.

Although, as mentioned earlier, because soft assets values are created synergistically with spill-over, and scaling-up effects, a true picture can only be mapped through actual experiences, observations, and implementation from key stakeholders. This was the basis on which the questionnaires were developed which highlighted city level priorities and approaches involving real assets, impact, and actual cases of value creation.

1. ***Maintaining a balance between physical infrastructure and soft (or intangible) assets in smart city development.*** The series of questions under this pillar sought to understand the views of city officials and other key stakeholders in considering

using soft assets, including 'successful action/benefit practice', 'how soft assets were conceived during planning and used during implementation, including the results, as well as the challenges and the overall benefits of using soft assets.
2. ***'Resilience' and the 'Covid-19 pandemic effect on incorporation of soft assets'*** were also included to understand how the city made decisions and took actions to prevent and prepare for disasters and the Covid-19 pandemic.
3. ***Value creation action and impact cases.*** Provide examples of decision-making to use soft assets to create value and mapping
4. ***Consultation process, quality of communications, and community engagement.*** The responses in this area are linked closely to the India's Ease of Living Index.
5. ***Review, feedback, and the learning and sharing process.*** The series of questions under this pillar were designed to examine feedback loops or any systematic processes used by the city officials to evaluate Smart City Mission implementation results in the context of their urban development strategies.

5.2 Case 1: Bhubaneswar

5.2.1 Desk Top Research (Proposal Stage)

Initial findings (Connectivity)

The Smart City Mission proposal is assessed using the qualitative identification framework tool. Table 5.2 presents a series of actions and initiatives categorized under each soft asset capital element (the top row, horizontal axis) and the nine value elements (the left column, vertical axis). In this context, the connections made on both axes are called 'links'. The 'link' refers to specific actions linked (or associated) with five soft assets within each of the nine elements. For example, 'planning and developing effective solid waste collection' is linked to both 'people/human capital' and 'institutional capital'.

The results of the Table 5.2, after applying the network analysis tool, are shown in Fig. 5.1.

Figure 5.1 shows the interconnection based on each value element (larger nodes are shown in purple and green) and actions and/or initiatives (in yellow, smaller nodes). Each line (edge) represents five relevant soft asset capital elements. This also demonstrated by the linked parameters in Figs. 5.2 and 5.3.

A total of 391 actions/initiatives are identified and distributed across the nine value elements and five soft asset capital elements. Here, each individual activity/initiative and the nine value elements associated with them are referred to as 'nodes'. This is because they serve as basic data essential for creating a visualized network. Since many of them are linked or shared with the same nine value elements and soft asset capital components, the non-cumulative number of nodes became 161 (that is the final number of nodes). Each line (edge) in the left hand column represents five

Table 5.2 Number of actions/initiatives (soft asset consideration): Bhubaneswar (proposal stage)

	People/ Human capital	Institutional capital	External capital	Technology (Information) capital	Social capital	Total	%
Resource Management	14	18	7	2	7	48	12.3
Accessibility (Mobility)	8	12	6	6	1	33	8.4
Health Well-being	16	18	7	4	5	50	12.8
Equal Access to Services	16	21	10	6	5	58	14.8
Smarter Services via Technology	9	19	11	22	6	67	17.1
Community Engagement	4	7	2	2	10	25	6.4
Quality Communications	5	9	4	9	10	37	9.5
Security	7	13	3	5	3	31	7.9
Resilience	14	15	5	4	4	42	10.7
Total	93	132	55	60	51	391	100.0
%	23.8	33.8	14.1	15.3	13.0	100	

Source Created by the author based on the desk top research, Wataya, Shaw (2022), Wataya (2023)

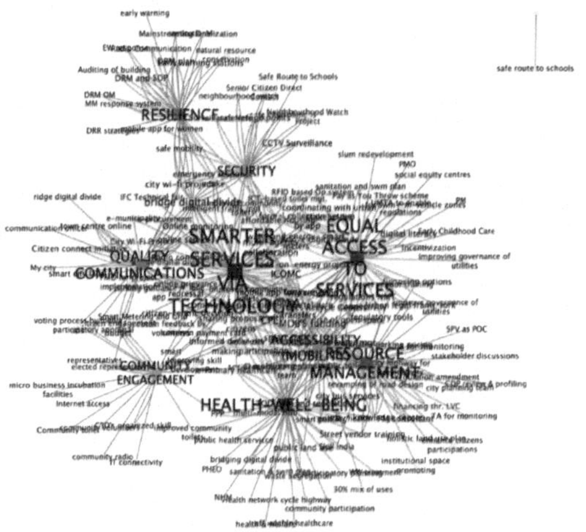

Bhubaneswar	Betweenness Centrality
SMARTER SERVICES VIA TECHNOLOGY	0.32951008
EQUAL ACCESS TO SERVICES	0.228870407
HEALTH-WELL-BEING	0.224834217
RESOURCE MANAGEMENT	0.222476692
RESILIENCE	0.162294326
QUALITY COMMUNICATIONS	0.135301145
COMMUNITY ENGAGEMENT	0.134712723
SECURITY	0.129385676
ACCESSIBILITY (MOBILITY)	0.125834117

Fig. 5.1 Bhubaneswar—Soft asset value elements, soft asset capitals, activities/initiative, and Betweenness Centrality (*Source* Wataya and Shaw 2022, Wataya 2023) (image and data via Cytoscape, https://cytoscape.org (accessed on 4 January 2022))

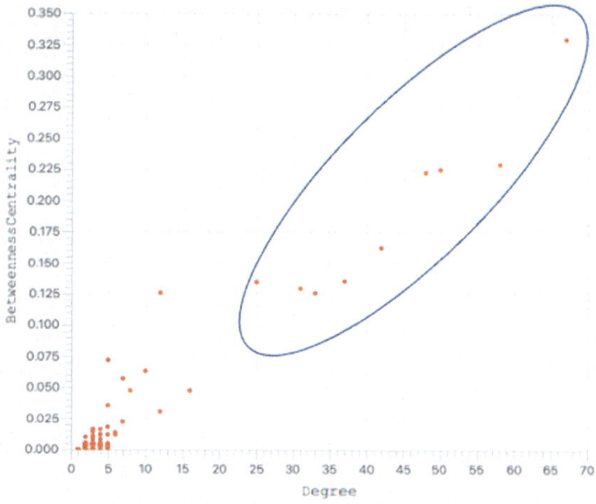

Fig. 5.2 Node degree distribution. *Source* Wataya and Shaw (2022), Wataya (2023)

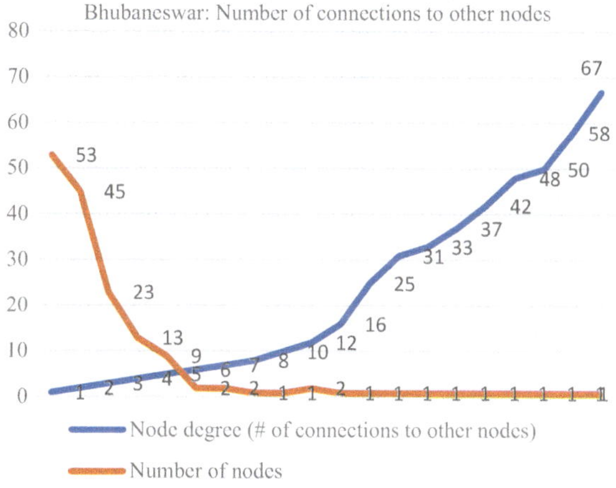

Fig. 5.3 Number of connections to other nodes. *Source* Wataya and Shaw (2022), Wataya (2023)

main soft asset capitals. Smarter service via technology, equal access to services, and health and well-being are major areas where soft asset components are actively considered at the proposal stage.

Referring to Fig. 5.1, these larger, central nodes represented with purple squares and green nodes are supported by many nodes represented in small yellow nodes. For this network, they are more significant and influential in creating values than the larger nodes as they receive more support from each of the smaller nodes than others. In the actual process, these small nodes are more important than the larger nodes as they receive cross support from each smaller node. This was verified by the Scaling law discussed in Chap. 1. This is also demonstrated in the matrices and linkages in Figs. 5.2 and 5.3 by different parameters.

Figures 5.2 and 5.3 show the previous values of Fig. 5.1 in a graph to check the elements with the most significant impact. Figure 5.2 shows the degree of node distribution of the network presented in Fig. 5.3. The concept is to observe the correlation of two parameters. The degree of distribution is the number of connections with other nodes. 'Betweenness Centrality' of a node reflects the amount of control that this node exerts over the interactions of other nodes in the network (Network analyze online help, https://med.bioinf.mpi-inf.mpg.de/netanalyzer/help/2.7/).

In Fig. 5.2, a larger number represents a hub (large) node. The nine dots inside the blue oval circle represent a hub (large) node, each of which is a nine-value element. Figure 5.3 shows similar information, but presented in a different format, highlighting a number of connections to other nodes. The figure at the extreme right with blue (67) and orange (1) lines represents: '1 large node (Smarter services via technology) with 67 small (yellow) node connections'. The figure on the left with blue and orange lines represents 53 small (yellow) nodes with only 1 node connection'. All nodes are related, including the 53 nodes, but nearer nodes are more related than distant ones. In this network context, all 53 nodes are related to one large node, and they are considered as closer and more related to others as demonstrated by Tobler's law.

The concept of "betweenness centrality" describes the importance of nodes and level of soft assets consideration in this specific network picture. It is important to understand that these larger nodes represent influential nodes. But this 'importance' or 'influence' shown in Fig. 21 is purely derived from network analysis results and may not necessarily reflect actual performance or influence. In short, the influential node in 'the network' was generated by computation not based on actual performance, quality of actions or value creation. This needs to be validated by research data, including the survey results and feedback from this research.

Having said that, using the network analysis is useful because:

(i) It involves the results of the desk top research, including which elements and soft asset capital components were incorporated and the associated intensity.
(ii) Small nodes are more important as they produce larger influential nodes with better performance and quality.
(iii) The network analysis of the 'proposal analysis stage' and 'implementation stage (interview stage') can demonstrate: (i) how soft assets are interconnected; (ii) the inability to trace each soft asset in the value creation process; and (iii) the importance of assessing value creation at the citywide level.

Some findings from Figs. 5.1 and 5.2 are as follows: 'smarter services via technology' (67), 'equal access to services' (58) 'health well-being' (50) and 'resource management' (48) are more influential in the interaction of other nodes than the other five elements, namely 'accessibility (mobility)' (33), 'security' (31), 'quality of communications' (37), 'resilience' (42), and'community engagement' (25). Again, they are the results from the network analysis and do not reflect actual actions, results, quality, assessment or judgment of stakeholders on the ground. They may not be consistent with actual cases. Nevertheless, the network analysis remains a valid entry point to understand and manage assets for value creation over the course of time.

5.2.2 *Survey and In-Person Interviews (Implementation Stage)*

This section briefly explains the results of the Survey Questionnaires. Since the questionnaires used open-ended questions, the author analyzed the responses, and categorized them with the relevant key value element and the soft asset capital if not apparent from the responses (Wataya 2023).

Overall (Summary)

For comparison purposes, after conducting both sets of interviews and the survey regarding the incorporation of soft assets at the implementation stage, the results are as follows:

Although the sample size of respondents in Table 5.3 is relatively small, the city officials interviewed provided valuable insights about their reasons for utilizing soft assets in city development projects. Also the responses clearly show the hierarchy of importance among the nine key value elements and the connections among the nodes or sub-components which influenced the city decision makers.

Table 5.3 Number of actions/initiatives (soft asset consideration): Bhubaneswar from survey results

	People/ Human capital	Institutional capital	External capital	Technology (Information) capital	Social capital	Total	%
Resource Management	8	8	4	2	4	26	6.3
Accessibility (Mobility)	5	5	2	3	2	17	4.1
Health Well-being	11	11	2	6	4	34	8.2
Equal Access to Services	19	22	4	18	5	68	16.5
Smarter Services via Technology	18	19	2	19	5	63	15.3
Community Engagement	14	16	8	7	9	54	13.1
Quality Communications	21	24	9	14	9	77	18.6
Security	7	8	1	8	1	25	6.1
Resilience	14	17	3	11	4	49	11.9
Total	117	130	35	88	43	413	100.0
%	28.3	31.5	8.5	21.3	10.4	100	

Source Created by the author based on the survey, Wataya (2023)

The results show that two related major nodes i.e. 'equal access to services' and 'smarter services via technology' outweigh the others in influence in this network (connection) context. In contrast to the proposal stage, the 'quality communications' node was the largest and the 'community engagement' node outweighed the 'Resilience' node. 'Equal Access to Services' also outweighed others at the proposal analysis stage. Since the Bhubaneswar Smart City Mission emphasizes citizen-centric aspects, these nodes support key value creation elements of soft assets. The respondents valued better government-community relations/communications as the most predominant.

The results also show the different levels of social attitudes about the importance of soft asset capital and the complexity of soft assets in city development. For example, 'people/human capital' and 'institutional capital' tend to be viewed by city officials as a key 'enabler' for planning and executing smart city initiatives and projects and play a larger role in decision-making. However, actual use in practice depends on the skill level, knowledge and capacity of the city officials which is often not within the control of their institution. Also, the body of knowledge belongs to individuals and it is used as institutional capital only when they are working within the system. Individual knowledge and capacity of officials are not necessarily uniform across the institution and often there is no systematic process to build cross institutional knowledge.

Many respondents placed a priority on the importance of people/human capital and institutional capital in considering the use of soft assets as they are essential assets to deliver services. In the interview and survey results, the degree to which social capital was taken into account by Bhubaneswar city can be contrasted with Nagpur. The importance of social capital was explicitly highlighted in some Bhubaneswar survey responses. Long-term development by helping citizens (females, youth, elderly, marginalized people) to build confidence, develop inclusive leadership skills was strongly emphasized. This is one of the successful examples of integrating soft assets to create value over the time, not only for creating new actions, but also being valuable community/social assets to make people more self-reliant. Survey answers indicated that consultation and 'community engagement' was one of the most successful examples of using soft assets to create value.

Influence of large nodes

In the smart city proposal stage, there are three key nodes associated with soft assets consideration: Smarter service via technology, Equal access to services', and 'health, well-being' all of which many small nodes are linked. The results are presented in Fig. 5.4.

The data also shows that 'quality communications' is a major node or influencing factor. As mentioned, the city places weight on 'citizen-centric' factors, also related to other major nodes. One example is the 'Art trail' initiative that connects 'Resource allocation', 'health, well-being', 'quality communications', and 'smarter services via technology'. Some are directly and indirectly linked creating a new asset value now and in the future. These large nodes are supported by many small nodes as shown in Fig. 5.4. This also demonstrated by the linked parameters in Figs. 5.5 and 5.6.

Bhubaneswar	Betweenness Centrality
Quality Communications	0.248263895
Resilience	0.215206789
Equal Access to Services	0.198547011
Smarter Services via Technology	0.145955146
Community Engagement	0.137339431
Health, Well-being	0.123240847
Resource Management	0.104050374
Accessibility (Mobility)	0.04284788
Security	0.024159017

Fig. 5.4 Bhubaneswar—Soft asset value elements, soft asset capitals, activities / initiative, and Betweenness Centrality (image and data via Cytoscape). *Source* Wataya (2023)

Fig. 5.5 Node degree distribution. *Source* Wataya (2023)

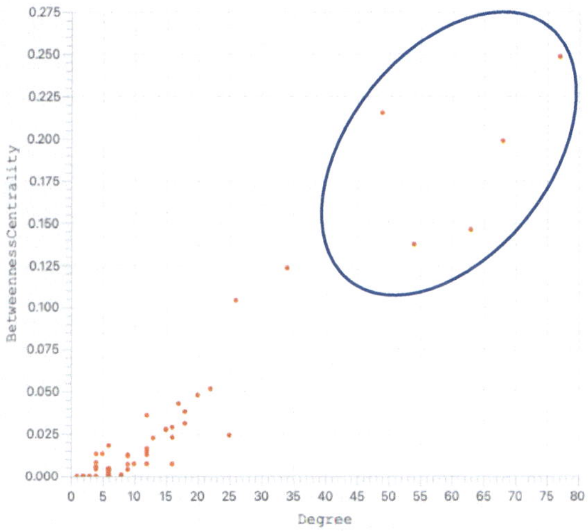

Fig. 5.6 Number of connections to other nodes. *Source* Wataya (2023)

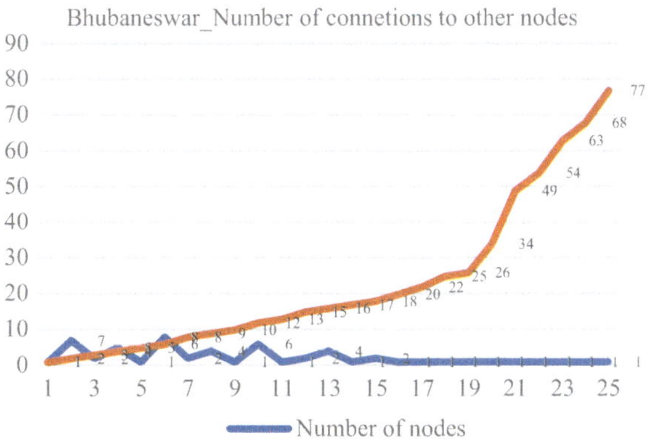

From Table 5.3 and Fig. 5.4, 'smarter services via technology' and 'equal services to all' play a major role in the planning proposal stage. However, as pointed out earlier (Sect. 5.2.1., this chapter) in regard to centrality, the performance of the larger nodes is different from the importance of the node in the network (connection). In Fig. 5.5, there are 4 dots in the top right circle. They represent 'quality communications' (Betweenness Centrality: 0.24826), 'equal access to services' (0.1985), 'smarter services via technology' (0.1459) but the second highest of Betweenness Centrality is 'resilience' (0.2152). Although the total result for 'resilience' is 49 (rank fifth among 9), it is the second most influential node in this network (connection). It means that the 'resilient' node plays a more crucial role in connecting different parts of the network. Also the 'resilience' node has a greater influence in shaping interactions among other nodes in the network.

The results of the network analysis can be verified by the EOL results of Bhubaneswar. It corresponds to the results where 'City resilience' and 'Housing and Shelter' outweigh the others. They correspond to 'resilience'; and 'equal access to serviccs' in this book where soft assets incorporation was more important than other elements. Also, given that some items outweigh other items in the interview stage, the network analysis results remain valid in terms of the influential node (Table 5.4).

Table 5.4 Proposal stage and interview stage comparison: Bhubaneswar

Bhubaneswar

	Proposal stage ranking		Number of actions/ initiatives	Betweenness centrality (BC)	Ranking (BC)
Total number of actions/initiatives	1	Smarter services via technology	67	0.329	1
	2	Equal access to services	58	0.228	2
	3	Health, Well-being	50	0.224	3
	In-person interview stage ranking		Number of actions/ initiatives	Between centrality (BC)	Ranking (BC)
	1	Quality communications	77	0.248	1
	2	Equal access to services	68	0.198	3
	3	Smarter services via technology	63	0.145	4
	4	Community engagement	54	0.137	5
	5	Resilience	49	0.215	2

Source Created by the author based on the survey, Wataya (2023)

Soft assets and resilience

Due to its geographical location and climate, Odisha State is prone to a variety of natural disasters such as cyclone, floods, tsunami, heatwaves and climate impacts. The impact of the disaster events on society and environment has become more complex as communities and urban areas grow and more advanced. Odisha State Disaster Management Authority (OSDMA) focuses on 'preparedness, capacity building & inculcating risk informed programming strategy in key departments in order to increase the resilience of state, departments and communities.' (OSDMA, https://www.osdma.org/disaster-risk-reduction/#gsc.tab=0). Bhubaneshwar has also been implementing climate risk management through disaster preparedness and mitigation under the Government of India and UNDP program, focusing on enhancing institutional capacities and integrating climate risk reduction measures in development programs, undertaking mitigation activities and enhancing community capacity (Bhubaneswar Municipal Corporation, https://www.bmc.gov.in/departments/disaster-management). Bhubaneswar placed priority on city administration and community collaboration preparedness and building resilience. The survey results showed broadly the same approach in using soft assets in each city currently and for the future.

As stated earlier, Bhubaneswar is frequently affected by cyclones due to its location on the coastal region of the Indian Ocean. The State of Odisha experienced more than 10 major cyclones since 1999. Recently Cyclone Amphan in 2020 devastated west Bengal and Odisha. The Bhubaneswar Municipal Corporation prepared its action plan focusing on preventing flooding and handling evacuations. The plan was designed to address road clearance, drainage and water supply, evacuation and rescue operations in coordination with various key city specialists such as engineers, field level officials, disaster response units, fire-fighting responders, and civic officials. During the disaster, all ward officers, tax collectors and other city field level officers of the city agencies were also requested to report causalities, areas of concern and issues that made it possible to closely monitor and take remedial actions in a timely manner (The New Indian Express).

Science-based preparedness is very important to obtain accurate data. Odisha is also vulnerable to the effects of extreme weather such as drought, floods, heatwaves and storms. After the most severe heatwaves and devastating cyclones, the government shifted (changed) its strategy from response to prevention. Then, it was further aided by the central Government making the disaster risk management approach more comprehensive, systemic and inclusive from the national to local level (Nature India).

Kawyitri and Shekhar (2021) assessed the vulnerability and capacity of Bhubaneswar using a case study of the impact of the Fani cyclone (2019) on the city. Sixty six vulnerability and capacity indicators were used to assess overall capacity. The study found a higher capacity to respond to the cyclone from the social and human dimension, but the response was less effective in the physical, smart-city, and financial areas. The city's improved preparedness was partly due to India's Smart Mission program which developed better infrastructure. At the time of the disaster, important operational action was taken to mitigate the impact of the damage. Programs under the Smart City Mission enabled administrators to provide social and health services. Kawyitri and Shekhar argued this type of program empowered a strong community response with citizens becoming more self-dependent and adaptive (Kwayitri and Shekhar 2021).

The improved preparedness of the cities is also supported in the survey results which demonstrated: (i) successful ways of incorporating soft assets to build resilience during smart city mission; and (ii) examples of soft assets being used to improve resilience. In Bhubaneswar, apart from the 'Resilience' indicators, the results also show that 'quality communications' are given more weight in building resilience. Results of interviews and surveys classified by the author using the same qualitative framework (45 items) are shown in Table 5.5.

Regarding successful ways to build resilience, most of the responses relate to communications, increasing understanding about the risks, such as community-led initiatives, gauging citizens awareness from their experiences and risk management actions, strong grievance redress mechanisms, proper protocol for disaster management and evaluation, emergency support functions, information dissemination, including mass-scale evaluation, well-connected administrative response effectively mobilized communities. Other areas are the effective use of natural resources for building city resilience against monsoons, floods, urban heating, but also provision for open space. Regarding soft assets required for improving resilience, most of the responses also referred to community interaction, youth activities, mobile applications for awareness raising and accessing basic services as well as use of technology among the community to address sanitation and other local requirements. Engaging volunteers to conduct door to door activities was an effective way to raise risk awareness in communities. Also mentioned were enhancement of knowledge and awareness on preparedness and mitigation measures, improvement of human, institutional and social aspects for long term resilience.

Overall,"equal access to services" and"quality communication" are more prominent and correspond with the survey results. Specifically, when examining the incorporation of soft assets, "external capital" and 'social capital' are also more prevalent, aligning with the city's emphasis on preparedness in parallel with hard infrastructure, indicating the development of a nurturing soft enabling environment in the resilient domain (Tables 5.5 and 5.6).

Table 5.5 Successful way incorporating soft assets

Bhubaneswar (Question 8–1)	People/ Human capital	Institutional capital	External capital	Technology (Information) capital	Social capital	Total
Resource Management						0
Accessibility (Mobility)						0
Health Well-being	1	1				2
Equal Access to Services	7	8		8		23
Smarter Services via Technology	5	6		6		17
Community Engagement	4	5	1	4	1	15
Quality Communications	7	11	2	8	1	29
Security	4	5		4	1	14
Resilience	13	16	3	10	4	46
Total	41	52	6	40	7	

Source Created by the author based on the survey, Wataya (2023)

Table 5.6 Three instances of soft (intangible) assets crucial for enhancing resilience

Bhubaneswar (Question 8–2)	People/ Human capital	Institutional capital	External capital	Technology (Information) capital	Social capital	Total
Resource Management			2	1	2	5
Accessibility (Mobility)						0
Health Well-being	1	2	1	1	1	6
Equal Access to Services						0
Smarter Services via Technology						0
Community Engagement	1		8	7	7	23
Quality Communications	1	2	10	7	9	29
Security	1	2	2	2	4	11
Resilience	4	6	11	8	11	40
Total	8	12	34	26	34	

Source Created by the author based on the survey, Wataya (2023)

5.2.3 *Value Creation Mapping*

Identifying cases illustrating value creation

Although the network analysis alone does not demonstrate the qualitative aspect of linkages or meaning of performance, this part needs to be informed by the in-person questionnaire responses. The questions about the successful cases of the impact of incorporating soft assets shows a multilayered, and holistic contribution to value creation.

Table 5.7 classifies the responses summarized in Table 5.8, using the qualitative framework employed in the network analysis. The classification also demonstrated that 'quality communications' and 'community engagement' are influential nodes that verify the findings of the connection or network analysis. It also confirms that Bhubaneswar's smart city vision and objectives incorporated soft assets effectively which created benefits for the citizens and the city.

The relationship between the five soft asset capitals and the value creation process

The illustrative cases obtained through interviews and survey questionnaires validate the potential value creation scenarios within the "Blackbox" concept detailed in Chap. 1. The key points about value creation include the relationship between inputs

Table 5.7 Classification of value creation actions: Bhubaneswar

	People/human	Institutional	Tech/information	External	Social	Total
Resource management		1			1	2
Accessibility (mobility)						0
Health, Well-being	2	4	2	2	3	13
Equal access to services	3	6	2	4	3	18
Smarter services via technology	2	5	3	2	2	14
Quality communications	8	10	1	8	8	35
Community engagement	9	10	1	8	9	37
Security	1	4	3	1	2	11
Resilience		1	1			2

Source Created by the author based on the survey, Wataya (2023)

Table 5.8 Some responses from interviewees regarding positive action and its impact (Bhubaneswar)

Positive action (1)	Phase-wise implementation **[BS, QC, CE x INST, HR, EX, SOCIAL]**	Self-defense and leadership training helped in building the confidence of the girls from the communities. The girls, who were earlier shy to even speak to other, have developed leadership skills and actively participated and even led many programs in the community and their colleges **[RM, CE, QC x SOCIAL, INST]**	During the planning of Art Trail, a 20–25 days long event envisaged to revive the overlooked and undermaintained (filthy) parts of the old city with rich cultural heritage and art elements, initial reluctance was observed both within the government and also among the citizens **H, WELL, EQ, QC, CE x HR, INST, EXT, SOCIAL]**	Institutional mechanisms like that of BUKC supports **[QC, CE, ACCESS, SMARTER x HR, INST, EXT]**	Unified city platform, namely Bhubaneswar.me was envisaged to bring together variety of existing municipal services, information and approvals offered by different municipal departments and agencies onto a single portal. This was planned to promote e-governance and enhance transparency by reducing the dependence on paper work **[SMART, H, WELL, QC, CE, SEC. x HR, INST, IT/INFO, SOCIAL]**	Similarly, initiatives such as street designing as people's space, bicycle sharing, art of city wherein visioning workshops were conducted with artists on what all can be done for enhancing artistic beauty and perception of the city gained popularity among the citizens and other stakeholders **CE x HR, INST, EXT, SOCIAL]**	Integrating soft aspects (i.e. citizen centric approaches) **[H WELL, SEC, BS x PPL/HR, INST, SOCIAL, EXT]**	

(continued)

Table 5.8 (continued)

Positive result/ Impact (1)	Phase-wise implementation was found to be a very impactful strategy. It helped in prioritizing the resources for applications which are targeted at critical services first, as identified through public consultations and studies	Self-defense and leadership training for the girls was found to be a very positive and successful action both in terms of cost–benefit analysis (financial implications) and the larger impacts it created in the community This initiative Fulfilled the envisaged objectives by igniting critical thinking process among the girls of the communities	After its launch the event was found to be a great success and attracted large participation of citizens which helped in transforming earlier neglected and filthy areas of old city into heritage areas; making them attractive and more usable	Bridging the communication gaps and materialize horizontal coordination among different smaller bodies and line departments associated with urban planning, development and provisioning of different urban services at the city level	The portal was found to be very effective during the COVID-19 pandemic when the physical accessibility to many basic services and municipal departments was restricted. The portal helped in supporting and linking the citizens with basic services including ration related services, recording of isolated persons, information dissemination on other micro services, etc	Bhubaneswar urban lectures involving best local and global professionals, key policy and decision makers, subject experts, etc. were conducted and were found to be more successful than envisaged	Encouraging more eyes and life on streets and in the city not only make the cities more lively but also provide a sense of security to communities, change the course of conversation towards positive and healthy choices, etc	Strong bureaucracy of the state was found to be effective

(continued)

Table 5.8 (continued)

Positive action (2)	Strong local support from the local corporators **[CE, QC, x HR, INST, EXT, SOCIAL]**	Addressing the minor needs of community of immediate nature **[CE, QC, SMART x HR, INST, EXT, SOCIAL]**	Involving directly with the community **[CE, QC, x HR, INST, EXT, SOCIAL]**	The telephone helpline 1929 was developed for grievance redressal of residents across the city **[SMART, ES, SEC, H Well, QC x INST, IT,]**		Using the soft capital of data and technologies **[SMART, SEC, RES, ACCESS x IT, INST]**	Consideration of social aspects right at the proposal stage **[CE, QC, x HR, INST, EXT, SOCIAL]**	
Positive Impact (2)	Closely working with the community and better understanding their needs and local context	Engaging with them to design long-term and sustainable solutions for aspects of urban development	Inculcating a sense of ownerships and initiating various community-initiatives, which ensures monitoring and sustainability of these actions	Beyond its envisaged role, the helpline was found to found to be very effective in providing tele-consultations and addressing the grievances of patients, migrants, etc. during the COVID-19 pandemic	Bhubaneswar Art Trail, an art excursion to regenerate infamous and unattended areas of the old city, was also found to have higher positive impact and acceptance than anticipated	Helping gather useful and large amount of environment-related data and information that were otherwise not available with authorities	Helping design solution most suited to the citizens instead of enforcing ready-made solution onto the communities	

(continued)

Table 5.8 (continued)

Negative action		Some initiatives failed to trigger the expected results and outcomes	It is observed that due to presence of overlapping authorities and prevailing inter-department coordination gaps and issues, it if often difficult to sustain some initiatives to their logical end and envisaged impacts, especially in the absence of a body like BUKC. BUKC currently is a body under BDA	Around 1600 Wi-Fi routers were installed across the city in 2016 to provide better internet access to residents of the city	Digital kiosks were planned and installed across the city to provide a digital access to residents not owning smart mobiles and other gadgets. These kiosks also have Emergency Call Button (ECB) that could be pressed by those in distress, for example, females facing any threat to their safety	The child friendly space guidelines are delayed due to bureaucratic reasons		

(continued)

Table 5.8 (continued)

Negative impact	Creation of dedicated parking slots was envisaged to make more usable space available to the residents of the city. The ideal and underutilized spaces thus created were then made available to the residents as clean and open access to parks, clear lanes, etc	These primarily included less active engagement of boys in activities envisaged at linking them with job opportunities, Open Defecation Free community related initiatives in a few pockets of the community and some initiatives on aspects of nutrition due to weak response from the NSS group of the college	However, it is perceived that it should be a permanent body at a state level or under an authority with bigger umbrella and powers of enforcements	Again, due to reduced internet cost and increased concern for data security, it is learnt that most residents were using their personal internet packages, leaving most of the routers underutilized. A way ahead to better use these routers now is possibly broadcasting of exclusive content through them	During planning stage, the growth in tele-density and heavy reduction in the internet costs could not be anticipated. This led to very lower usage and footfall in these installed digital kiosks making them redundant. Due to underutilization, this could be counted as wastage of existing resources	It was found to be inactive and unutilized		

Source Created by the author based on the survey, Wataya (2023)

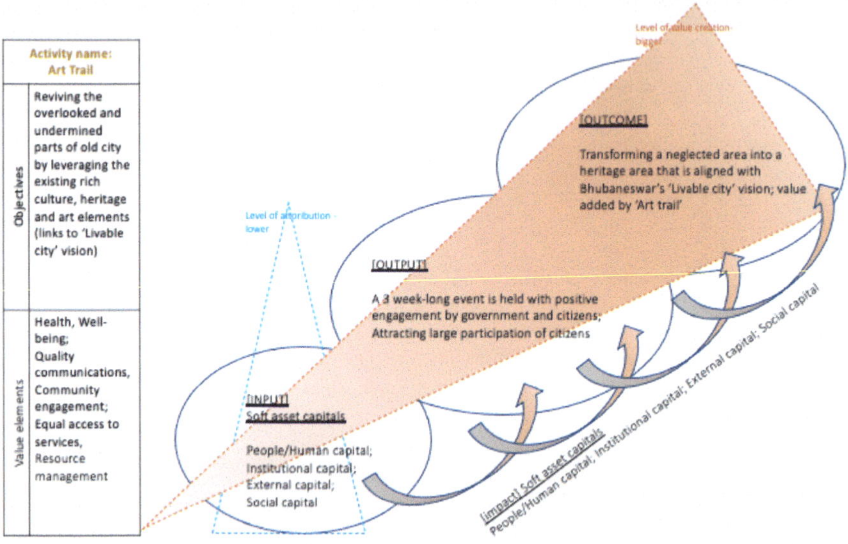

Fig. 5.7 'Value created' case (*Source* Created by the author based on the survey, Wataya 2023)

and outputs and depending on the purpose, become increasingly unclear over time. Conversely, value tends to become more likely as time passes. This applies both to positive and negative examples of value creation.

While there are many projects planned and implemented over time, it is not realistic to trace all projects reviewed at the proposal stage. Table 5.8 sets out some specific examples of how soft assets applications are considered and utilized in actual planning initiatives with inputs, outputs and outcome flows. Two case examples of a simplified value creation process are illustrated in Figs. 5.7 and 5.8.

(a) *'Value created' case (positive case)*

- Activity/project: Art trail
- Objectives:
  - Reviving the overlooked and neglected parts of the old city by leveraging the existing rich culture, heritage and art elements (links to 'Livable city' vision)
- Inputs:
  - *Soft asset capital*: People/Human capital; Institutional capital; External capital; Social capital
  - *Value elements:* Health, Well-being; Quality communications, Community engagement; Equal access to services

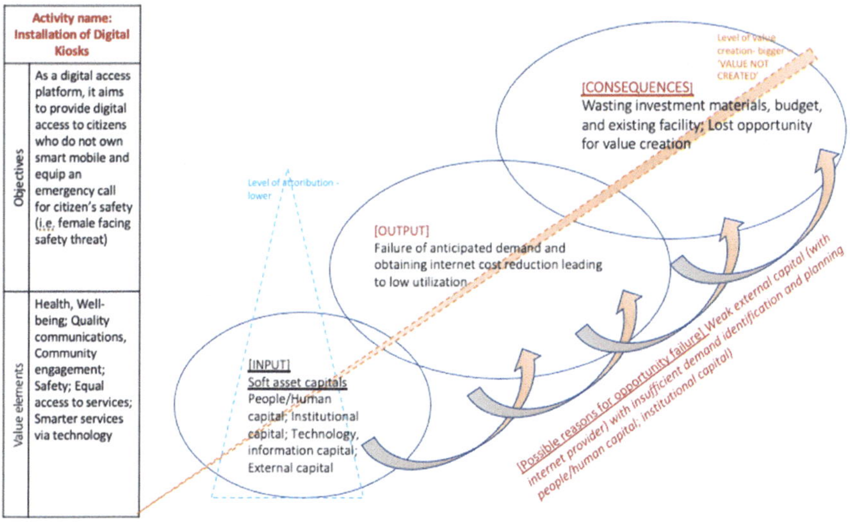

Fig. 5.8 'Value not created' case (*Source* Created by the author based on the survey, Wataya 2023)

- Output:
  - A three week-long event is held with positive engagement by city authorities and citizens; attracting a large citizen participation
- Outcome & future impact:
  - Transforming a neglected area into a heritage site that is aligned with Bhubaneswar's 'Livable city' vision, with an 'Art trail' adding value
  - *Soft asset capital*: People/Human capital; Institutional capital; External capital; Social capital
  - *Anticipated Value elements:* Health, Well-being; Quality communications, Community engagement; Effective resource management

'Value not created' can also provide answers to determine how to achieve a positive impact when the causes of the negative results are analyzed and appropriately used to identify and strengthen enabling factors underlying the incorporation of soft assets. This requires a systematic monitoring, evaluation and feedback process in the planning and development stage (Fig. 5.8).

(b) *'Value not created' case (negative case)*

- Activity/project: Installation of Digital Kiosks
- Objectives:
  - A digital access platform aims to provide digital access to citizens not owning smart mobile technology and permit emergency calls for a citizen's safety (i.e. female facing safety threat)

- Inputs:
  - *Soft asset capital*: 'people/human capital; 'institutional capital'; 'technology, information capital'; 'external capital'
  - *Value elements:* 'health, well-being'; 'quality communications', 'community engagement'; 'safety'; 'equal access to services'; 'smarter services via technology'
- Output:
  - Failure of anticipated demand and obtaining internet cost reduction leading to low utilization
- Consequences:
  - Wasting investment materials, budget, and loss of existing facilities; Lost opportunity for value creation
- Possible reasons for opportunity failure:
  - *Soft asset capital: weak external capital (with internet provider) with insufficient demand identification and planning people/human capital; institutional capital)*

Value creation mapping (canvas)

Although aspects of the value creation process remains a Blackbox, the survey responses provided greater insights into how values are created over time using a simplified visual tool.

The visual creation map or canvas below (Fig. 5.9) was referred to in a similar study by Pagano and Neubert (2015). This study focused on shareholder value creation using offshoring and outsourcing within the context of worldwide supply chains. They use the term 'a fuzzy box' concept (similar to the 'Blackbox' concept in this research) to describe stakeholders' processes used to achieve objectives. To understand how companies create value through a multi-perspective and multi-stakeholder model, this research developed a visual canvas approach. Below is a modified version of this approach: (i) the soft assets consideration value creation process mapping; and (ii) adaption from current Smart City Mission's program to future programs or initiatives.

The map created based on the Table 5.8 shows both positive and negative actions taken within the smart city mission. The positive action made impacts on both or either: (i) created assets; (b) made positive change; and/or (c) created value. Therefore, each one of the activities and initiatives (example in Figs. 5.7 and 5.8) were consolidated at the city level. There are also negative cases like some 'created assets' are expected to produce new value creation in any future projects or initiatives, leading to new action taken or changes created. In the case of negative actions with negative results, the opportunity for value creation was lost.

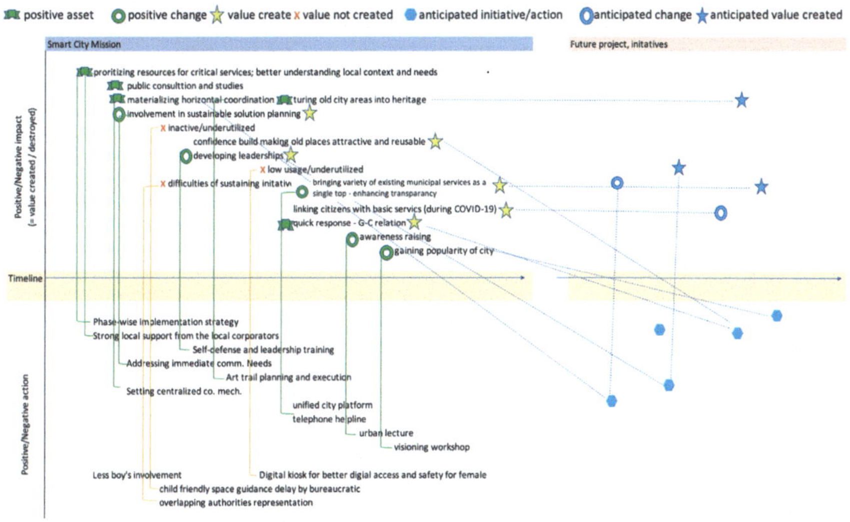

Fig. 5.9 Value creation mapping (Bhubaneswar) (*Source* Created by the author based on the survey, Wataya 2023)

Findings and conclusions from this visual mapping (canvas) are as follow:

- Action, impact, asset creation, changes, and value creation occurs throughout the project timeline. Inputs (action) and outputs or results (impact) are sometimes linear over the short term, but are more complex in the longer term as the unique nature of soft assets can result in spill-over, scaling up, and synergetic effects.
- Although the information and linkages are already understood, the visualizing process can help key stakeholders understand the value creation process that may not explicitly be discussed and shared with other stakeholders and decision makers.
- The value mapping needs to be refined further with inputs from more stakeholders, including focus group discussions to share their understanding to verify the value creation process and follow-up action. By doing so, a more detailed timeline and other key factors can be incorporated. Such exercises are useful for city decision makers to identify key resource capital and manage soft assets holistically within their overall development strategy.

From these results, key points for the review by city decision makers and stakeholders are to analyze what assets, actions and values are created and potentially how they can be used in the next stages of development (initiatives) or projects. The value creation map (canvas) demonstrates: (i) the various types of soft asset capital, related actions, and links to positive actions/valuc creation process; and (ii) impacts identified in the larger nodes that were examined previous section.

5.2.4 Consultation, Quality Communications, Community Engagement, and the "Learning and Share' Process

In the 'Ease of Living Index" (Chap. 4), the Citizen Perception Survey score is a key metric which measures the level of citizen satisfaction with city services. Stakeholder consultation is one of the best ways to gauge a citizen's immediate needs and determine the long-term engagement essential for optimizing the use of city services to improve their living environment.

5.2.4.1 Consultation Process, Quality Communications, and Community Engagement

In the survey responses, Bhubaneswar city interviewees highly valued the importance of consultation to ensure inclusive approaches from the planning, implementation and feedback review. As its Smart City Mission proposal stated, there was extensive public consultation involving five steps namely 'inform', 'consult', 'involve', 'collaborate' and 'empower' and the steps were carried out in three stages i.e. (i) options identification; (ii) prioritization; and (iii) planning for proposal preparation. The value of this consultation approach can also be observed during the implementation phase.

Consultation is one of the most widely accepted approaches for engaging communities and stakeholders, and strongly contributes to all nine key soft asset value creation elements. Also the interview data showed that both basic infrastructure services and some critical social issues were addressed by city administrators through detailed and well-targeted consultations.

(a) **Entry points (target audience groups):** The first question is to understand the types of consultations the city conducted. Different types and levels of consultation were provided with different entry points stakeholders and target audience groups. Table 5.9 represents a snapshot of these approaches. Establishing relationships with NGOs and grass root community groups are key and here the aspect of the soft asset 'external capital' is fully utilized to obtain inputs from communities. In addition, there are community level exercises that seek citizen feedback and suggestions, as well as making them aware of the community level outreach. Such engagement and consultations are crucial for building ownership of local projects among the local communities.

(b) **Effectiveness of Consultations**

Effectiveness of the consultation process is achieved by encouraging inclusiveness, empowering marginalized groups and nurturing local ownership. Some examples were shared by respondents about how these consultations were used to address immediate needs of communities. Consultations and engagement also helped to

Table 5.9 Process of consultation (implementation)

Purpose	Description	Entry point / stakeholders / target audience groups
Specific project focused	Varied consultations and human-centric studies were undertaken to better understand the needs and contexts of the local areas and communities and to design tailor-made and inclusive technological interventions for them. Some examples include: – Analysis of demographics on mobility (bus) to plan the number of seats – Smart Parking – Utilizing technology to reach out to the communities, on sensitizing them to necessary government schemes and services, and using social media for connecting with communities	
Consultation, monitoring and follow-up	Local NGOs (on Youth and gender) such as Humara Bachpan Trust (HBT) were engaged to make an entry into the community. This was followed by varied rapport-building activities including regular visits to the community, meeting with community leaders, and informal meeting with community members to identify issues, including the minor ones requiring immediate attention	NGOS
Strengthened G to C relation	Meeting with local government functionaries from the community such as ward officers, grassroot ANMs, ASHA, Aaganwadi workers, etc. were conducted to understand and strengthen the community-government linkage. This approach is effective for entering, understanding, and working closely with the community	Ward officers, grass root group, Area leaders/representatives
Community engagement for long term urban solutions	Community advisory meetings were conducted to address minor issues of an immediate nature. This approach also facilitated rapport building and gaining the confidence of the community, which is required for engaging with them in implementing long-term urban development solutions	Community advisory group
Engagement in social community mapping	The local youth groups were involved, and from among these groups, peer leaders were identified and further engaged in social and community mapping. The social and community mapping helped understand the key issues faced by the community in detail and also prioritized them to suit the community's need	Local youth groups

Source Created by the author based on the survey, Wataya (2023)

facilitate more effective integration of technology, policy and planning and dissemination of information and support through the community network. Specific examples provided by the respondents on the effectiveness of consultations are set out in Table 5.10.

(C): **Successes and Challenges:** The major lessons and successes of consultations listed by respondents are set out in Table 5.11. Respondents also referred to challenges. For example, although effective outreach with social media and technology is key, not all citizens have access to these options. This creates potentially negative impacts on citizens lives or a lack of understanding about the anticipated changes that could affect them. In this regard, communities are often reluctant to support the city's interventions. From the city's side, engaging communities to build trust can be time consuming and considerable effort is required to build awareness of the issues and outlining relevant solutions. Establishing trust, awareness and support between city authorities and

Table 5.10 Effectiveness of various approaches

Approaches	Effectiveness
Engagement and informal meetings	The existing issue of open defecation was addressed. The existing scheme of Open Defecation Free (ODF) was linked to aspects of women's dignity and the health and well-being of children and family members with the help of identified peer leaders and community volunteers
Digital support to community	Provided to generate awareness of local difficulties and challenges
Skill development and livelihood linkages	In support of UNICEF and UNFPA, training modules were delivered to youth groups. The trained youth were also linked to the Skill Development Institute and teams of relevant service providers. They were provided orientation on existing schemes and programs targeted for youth groups
Self-defense training were provided to youth groups and young girls	Boosted self-confidence of girls of the community for participating in other ongoing activities and meetings in the community
Youth groups were also sensitized on aspects of substance abuse	Gender-based equality and violence, as well as environment protection were the focus. The relationship nurtured with the youth groups also helped in organizing different college-led initiatives such as Youth for Smart City
Different days of importance were also observed in the community along with active participation of the community members	Youth groups, peer leaders, etc. to generate awareness on these crucial topics Efforts were also made to mainstream the sanitary workers in the community

Source Created by the author based on the survey, Wataya (2023)

communities can be challenging and requires a significant investment of time and effort by both sides.

Table 5.11 Successes and challenges

Factors	Actions	Comments
Frequency	Large number of events for engaging with communities	
Conscious effort to engage community	Strong community engagement	Active involvement of citizens and their representatives in the implementation of projects. This ensures that citizens are aware of reasons why the project is planned and how it is being implemented and how it would benefit the citizens
Diverse communications channels—mobile applications	Mobile-based application for linking the community with existing government services and community engagement High reach out to schools and colleges for ensuring higher participation of students, teachers, and communities	There is a strong presence of local authorities on social media due to which they are easily accessible and also very responsive to citizens needs and feedback It was also monitored by recording the number of respondents and number of viewers/responses on audio-visual materials Support by the BMC and BSCL during the COVID-19 pandemic
Feedback, Grievance redressal mechanisms	Strong mechanisms for seeking feedback of the citizens and using them to improve the initiatives and their implementation Responsiveness of officials to complaints and grievances	Communities are aware of and have witnessed local authorities promptly responding to their grievances. Even the post-disaster restoration of basic services is prompt, increasing the confidence of the communities in the local administration and authorities
	Strong two-way communication between citizens and authorities	
Incentivizing citizens	Provisions such as perks to travel abroad, exposure to learning programs, etc. were also planned and conceptualized to enhance performance of officials and motivate them	

Source Created by the author based on the survey, Wataya (2023)

5.2.4.2 The "Learn and Share" Process

The 'Making a City Smart' report (MoHUA, 2021b) sets out a recommended roadmap to establish a smart city. In the diagram of the step by step approach in the report, the 'Learn and Share' process (comprising comprehensive monitoring and evaluation) is emphasized as an important part of the process to establish a smart city.

The 'Learn and Share' process can be applicable, not only to planning stage, but also the whole development cycle. It is an effective way to strategically incorporate and embed soft assets at different levels of project initiatives and activities from policy, strategy, and project operations. To better understand the different approaches in the city planning and implementation stage, the four following specific questions were asked in the survey: (a) the process undertaken; (b) describing actual cases; (c) challenges to materialize "Learn and Share"; and (d) evaluation of vision, goals, actual activities and the soft assets monitoring process.

These results closely relate to the different intangible asset management and knowledge approaches explained in Chap. 3. Soft asset capital provides a unique contribution to city innovation and values. To understand these unique values, it is important to review how the 'Learn and Share' process is operationalized, including learning from experiences, successes, failures, and organizational internalization. Then they are fed into the next project or action design and formulation through a feedback loop.

(a) **The Learn and share process in operations**

Bhubaneswar city uses the Ministry of Housing and Urban Affairs (MOHUA)'s system and other systems at national level. The Special Purpose Vehicle (SPV) uses information (command and control center) to analyze cases for learn and share. In addition to the SPV, there is also the Bhubaneswar Urban Knowledge Centre (BUKC), a centralized coordinating functionwhich plays a key role of central coordination role to implement the Smart City Mission processes. The BUKC also incorporates an internal learn and share mechanism producing key materials and documentation to build institutional capacity. Associated survey results are set out in Table 5.12.

When the SPV eventually closes in Bhubaneswar, the user departments will assume responsibility and ownership. Currently, the BUKC supports communication and coordination among key departments implementing smart city proposals. BUKC systematically strengthens soft capacity by providing knowledge, learning and capacity building extending the benefits of the Smart City Mission in the long-term.

(b) **Challenges in establishing the "Learn and share" process**

Perhaps one of the most important challenges is maintaining the centralized function at local city level after the end of the Smart City Mission program. Although the centralized organization like the BUKC is key to undertaking coordination across different departments, the roles and differences from the Command Center is not clear. Clear roles and responsibilities are vital for an effective centralized system and

Table 5.12 "Learn and Share" process undertaken

Level	Entity, process	
Central	Primarily data sharing and exchange are conducted using the MoHUA's Data Exchange Unit (portal on data sharing)	It is noted that the portal's emphasis is more on quantity (numbers) while quality as a factor seems to be ignored a bit
National	One of the BUKC's core function. Supporting coordination, knowledge and information sharing among the relevant authorities and consultants	Internal knowledge session for learning and sharing
	BUKC undertakes extensive research on each of its verticals and study the existing and evolving trends, good practices, reasons for their success and how can they be best adopted in Bhubaneswar (from all possible lenses of finance, human resources, administration, etc.)	Followed this process, BUKC supports the development of informed and contextualized proposals for introducing new initiatives or projects in the city
	Smart city was conceptualized to have a unified system, namely Command and Control Centre which was envisaged to bring together different departments, agencies and stakeholders for close coordination and information sharing	
Learning network	Bhubaneswar is part of many national and international networks	International Smart City Network (ISCN), World Urban Forum
	Besides, efforts are undertaken to study and learn how other cities are utilizing and validating the huge data and information collected through different tools and technologies such as CCTV, RFID, etc	
	BSCL has been pioneer in effectively using data through scenario-based planning to study different cases	
Dissemination	BUKC is development of the project brochures documenting key features of different planning and implementation initiatives undertaken by the BUKC	This helps in developing administrative and institutional memory for years to come which helps in ensuring sustainability of initiatives despite change in administrative and technical heads and human resources

Source Created by the author based on the survey, Wataya (2023)

it also provides a clear path to develop an exit strategy at the end of the Smart City Mission (Table 5.13).

One clear example is the grievance mechanisms supplied with data and information from multiple sources collected at the Command and Control Centre along with the social media handled by BUKC and other relevant agencies such as Bhubaneswar

Table 5.13 Cases and challenges of learn and share

Good case	The Integrated Command and Control Centre, which was envisaged to promote inter-department coordination and knowledge sharing, needs to be effectively utilized for further improving the learn and share process
	A good practice followed in Bhubaneswar currently is that all BMC, BDA, BSCL is administratively led by a single official which makes it easier to integrate actions
	This integration can further be leveraged in ensuring mutual learning and sharing with other line departments and stakeholders too
Challenges	The administrative linkages and coordination among the three principal authorities, namely, Bhubaneswar Municipal Corporation (BMC), Bhubaneswar Smart City Limited (BSCL) and Bhubaneswar Development Authority (BDA) are strong. Often the same official is heading all three administratively. However, the linkages and integration with other line departments and stakeholders may further be improved for better learning and sharing
	The challenges of duplicity in some initiatives of varied departments such as surveillance systems, etc. could be avoided or streamlined with better coordination and information sharing among different planning and implementation departments and agencies
	Due to frequent changes in bureaucracy of different key departments and agencies associated with Smart City planning and implementation, communication often becomes challenging without having a robust mechanism of creating institutional memory
	Absence of inter-operability of monitoring systems of different departments and agencies also act as a hindrance for integrated monitoring of different services of the city

Source Created by the author based on the survey, Wataya (2023)

Smart City Ltd. (BSCL). The grievance redress mechanism is backed by strong procedures (SOPs) to ensure timely response and action by the relevant agencies. The process helps to build trust with citizens. The robust grievance mechanism is also used to regularly capture the feedback from the end users and communities to improve service delivery.

(c) **Evaluation of the vision, goals, activities and monitoring of soft assets**

The Smart City Mission monitoring process explained in Chap. 4, comprises four levels, including a Monitoring Committee consisting of the Commissioner or Deputy Commissioner (BMC), CEO (BSCL), Vice-chairman (BDA) and other officials who monitor and evaluate monthly updates and progress of the mission.

A major challenge is the evaluation of the vision and goals. Developing a baseline is crucial for evaluating the effectiveness and progress of the implementation and value impact of the smart city mission initiatives. At the proposals stage, key performance indicators (KPIs) are developed. From the survey results, the KPIs are benchmarked against overall sound practices in India and abroad. However, within the consultations for this research, the city's evaluation remains internal and unpublished so that details are hard to review. This highlights challenges in capturing the impact of the city intervention in the short term, while also revealing the absence of a systematic qualitative evaluation approach for managing soft assets. In this context, two indexes, namely the Ease of Living (EOL) and the Municipal Performance Index (MPI) can provide some answers. As mentioned in Chap. 4, the citizen-based perspective survey which assesses different project phases is also one of the most useful resources for evaluating smart city development.

5.3 Case 2: Nagpur

5.3.1 Desk Top Research (Proposal Stage)

Initial findings (Connectivity)

Applying the same analytical approach to the case of Bhubaneswar, the following is a set of data demonstrating Nagpur's connectivity. Table 5.14 shows that a total of 356 actions/initiatives were identified, but many are shared among nine value elements and five soft asset capitals. Hence, the final number of nodes became 136. To demonstrate their interlinked relationships, they are pictured in Fig. 5.10.

Table 5.14 Number of actions/initiatives (soft asset consideration): Nagpur (proposal stage)

	People/ Human capital	Institutional capital	External capital	Technology (Information) capital	Social capital	Total	%
Resource Management	10	10	3	2	2	27	7.6
Accessibility (Mobility)	10	8	3	5	3	29	8.1
Health Well-being	15	20	4	5	8	52	14.6
Equal Access to Services	19	15	4	12	7	57	16.0
Smarter Services via Technology	19	18	4	18	8	67	18.8
Community Engagement	2	2	2	3	7	16	4.5
Quality Communications	3	4	2	7	6	22	6.2
Security	12	13	7	7	5	44	12.4
Resilience	12	19	4	4	3	42	11.8
	102	109	33	63	49	356	100.0
%	28.7	30.6	9.3	17.7	13.8	100.0	

Source Created by the author based on the desk top research, Wataya, Shaw (2022), Wataya (2023)

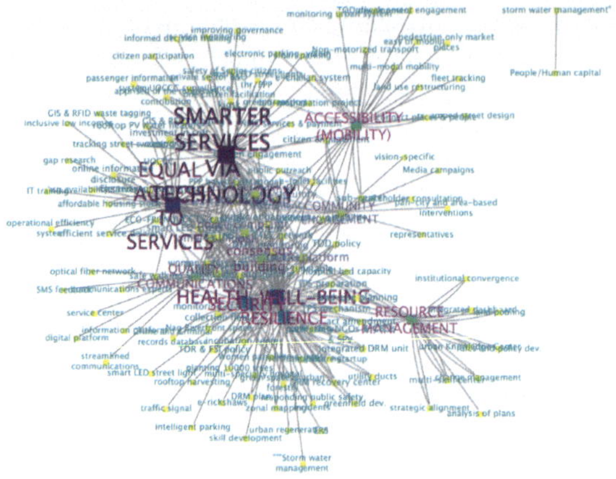

Nagpur	Betweenness Centrality
SMARTER SERVICES VIA TECHNOLOGY	0.326852456
HEALTH-WELL-BEING	0.272815576
EQUAL ACCESS TO SERVICES	0.187625183
ACCESSIBILITY (MOBILITY)	0.17629082
SECURITY	0.164037804
RESILIENCE	0.144209377
RESOURCE MANAGEMENT	0.134545444
Consensus building	0.126544333
QUALITY COMMUNICATIONS	0.12293745
Use of a statutory instrument	0.082073171
COMMUNITY ENGAGEMENT	0.080847656

Fig. 5.10 Nagpur—Soft asset value elements, soft asset capitals, activities/initiative, and Betweenness Centrality (image and data via Cytoscape) (*Source* Wataya and Shaw 2022; Wataya 2023)

The indicators are potentially exclusive, but as mentioned in Chap. 3, the qualitative tool has some limitations due to its objective and the classified types of the nine parameters ("nine value elements") and the five dimensions ("soft assets capitals"). The 'smarter services via technology' (coming within the nine value elements) and 'technology and information capitals' (under the five soft asset capitals) are differentiated by 'element' and 'soft asset capitals' with some overlapping between them. Since the use of technology is a fundamental factor in building smart and resilient cities, the 'smarter service via technology' element should be included as one of the value creation elements, while 'technology, information' capital remains as one of the five soft asset capitals.

Nagpur's network visual image appears denser than that of Bhubaneswar. Based on the data used for this network visualization, the Cytoscape analyzer function computed that the network density analysis figures are 0.025 for Nagpur and 0.019 for Bhubaneswar. The figure range is 0 to 1 with 1 the highest density node connectivity and the highest velocity of spreading information within the network. In this context, Nagpur's case is slightly more influential among connected nodes than Bhubaneswar, but both density figures are very low, thus the difference between them is unknown. Also this is only the network analysis level and does not reflect real performance or results (Wataya and Shaw 2022; Wataya 2023).

Same as Bhubaneswar, Figs. 5.11 and 5.12 shows the top three large nodes of 'Smarter service via technology' (67), 'Equal access to services' (57), and 'health well-being' (52), followed by 'security' (44) and 'resilience' (42), are highly influential in the proposal in terms of soft asset consideration. Nagpur also shows consideration of 'accessibility' (29) and 'resource management' (27) that are directly aligned to the city's vision (Wataya and Shaw 2022; Wataya 2023).

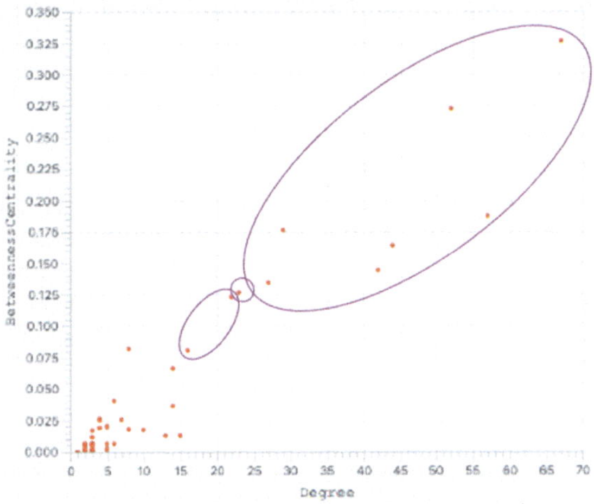

Fig. 5.11 Node degree distribution (*Source* Wataya and Shaw 2022; Wataya 2023)

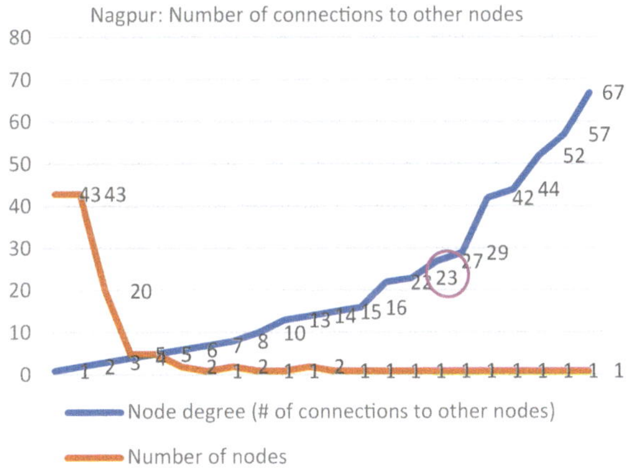

Fig. 5.12 Number of connections to other nodes (*Source* Wataya and Shaw 2022; Wataya 2023)

Unlike Bhubaneswar, Fig. 5.11 also shows 'consensus building' as a relatively influential node (a dot with degree scale '23' in Figs. 5.11 and 5.12). It is attributed to actions and/or initiatives, not associated with the nine value elements. However, 'consensus building' was frequently identified across the nine elements and the five soft asset capitals. That appears to make it more influential than other small yellow action nodes. Additional research to observe changes could be a valuable next inquiry.

5.3.2 *Survey and In-Person Interview (Implementation Stage)*

The same approaches for Table 5.14 were applied to prepare Table 5.15. Overall, compared to Bhubaneswar, there was less information from the Nagpur interviewees. It may be difficult to understand some terminologies in the questionnaires so some time was needed to ensure fuller understanding by the respondents. Alternatively, there may have been limited interview time due to competing priorities. Also, the arrangements for contacting appropriate respondents was different from the Bhubaneswar process. Nevertheless, the different affiliations, including NSSCDCL (SPV) and others provided useful responses.

Overall findings

The survey interviews were conducted selectively and targeted different respondents (SPV, city government units, private agencies, NGOs, academia). As mentioned in the Bhubaneswar section, the total number of responses is a useful reference to understand the areas and types of soft asset incorporation and important nodes or value creation areas in this network (or connection). Also there is a hierarchy among the nine key soft assets key value elements. The network was generated by

Table 5.15 Number of actions/initiatives (soft asset consideration): Nagpur from survey results

	People/Human capital	Institutional capital	External capital	Technology (Information) capital	Social capital	Total	%
Resource Management	14	14	3	12	1	44	17.1
Accessibility (Mobility)	3	3	0	3	0	9	3.5
Health Well-being	10	10	2	10	0	32	12.5
Equal Access to Services	13	13	1	15	2	44	17.1
Smarter Services via Technology	13	13	1	17	1	45	17.5
Community Engagement	3	3	2	1	1	10	3.9
Quality Communications	7	7	3	4	3	24	9.3
Security	7	5	1	7	1	21	8.2
Resilience	7	7	3	7	2	26	10.9
	77	75	16	76	11	255	100.0
%	30.2	29.4	6.3	29.8	4.3	100.0	

Source Created by the author based on the survey, Wataya (2023)

computations not based on actual performance, quality of actions or value creation, but is nonetheless useful to compare to 'proposal stage' and to validate with results and feedback.

The results show that the similar major nodes i.e. 'smarter services via technology' and 'equal access to services' are more significant in this network context. Since Nagpur's smart city mission strategy electronically connects people with the city government to create an inclusive eco-system, the efforts to utilize resources are aligned with the overall city's overall vision. As for 'resource management', it is a cross-cutting value creation area, but 'technology/information capital' is rated highest. The network (or connection) tables and the figures show that incorporation of the soft assets is effectively aligned with the city's vision and objectives.

From the survey feedback, the key benefit of using soft assets show significant potential to developing individual skills, gaining knowledge, making future behavioral changes in career development and more effectively addressing projects. The use of smart solutions/e-governance to serve the citizens, conduct consultations and engaging citizens to better understand bottlenecks and gaps on the ground has strong potential. Enabling scientific and informed decision-making and potential upscaling of the current practices were also raised in the survey. It was a reliable feedback from different stakeholders i.e., local authority, SPV, NMC and private sectors. Although the responses are unique coming from respondents with different roles and responsibilities, they generally show similar positive perspectives on soft assets and greater efforts in the future to incorporate them in their work.

However, respondents share common issues such as an imbalance between administrative policies, required resource availability, delivery time schedule, and readiness to accept new administrative projects/processes. This relates to other feedback on ad hoc approaches of considering and using soft assets. Furthermore the use of soft assets are often restrained by a lack of political support and other city priorities. Additionally, the human capital aspect is often implicitly and not specifically considered in the planning and implementation process. Hence, capacity and skills deficits of city administrators, coupled with insufficient attention to operational levels and appropriate staffing, remain unaddressed.

On the other hand, respondents also referred to specific opportunities for using soft assets at different levels because of the influence of global practices and initiatives and directives at the national level; for example, the recent focus on e-mobility and related asset creation. Also at the project implementation level, service providers and contractors were expected to add value by contributing to value creation from soft assets perspectives, as they have specific terms of reference for projects and possess broader professional experiences to identify, incorporate and use soft assets in the planning and implementation stages.

Influence of large nodes

Figure 5.13 shows the influential nodes in this network (connection) based on the survey results. Both "Smarter service via technology" and "Equal access to services" are linked to many small nodes. This is further illustrated by the connected parameters presented in Figs. 5.14 and 5.15.

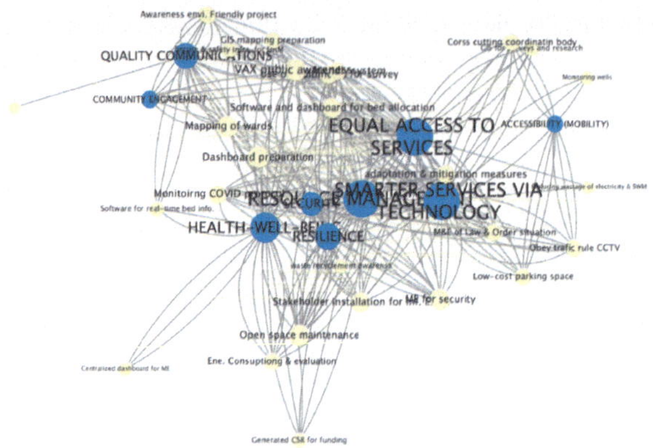

Nagpur	Betweenness Centrality
Smarter Services via Technology	0.245867885
Equal Access to Services	0.233814169
Health, Well-being	0.181138548
Resource Management	0.121682687
Quality Communications	0.113916912
Resilience	0.084899216
Security	0.055609655
Community Engagement	0.005294467
Accessibility (Mobility)	0.003154613

Fig. 5.13 Nagpur—Soft asset value elements, soft asset capitals, activities / initiative, and Betweenness Centrality (image and data via Cytoscape) (*Source* Wataya 2023)

Fig. 5.14 Node degree distribution (Wataya and Shaw 2022)

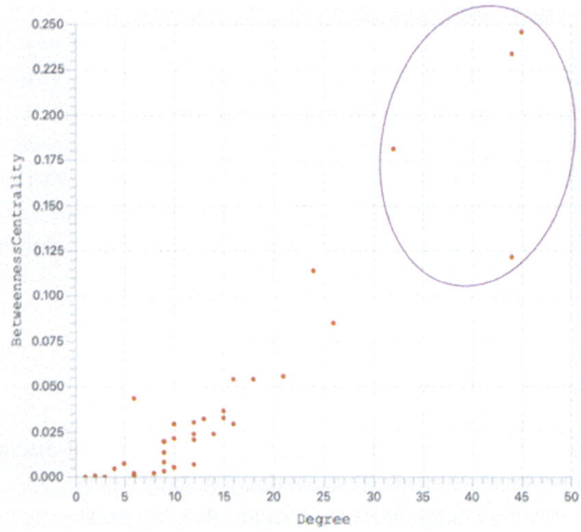

Figure 5.14 shows that there are 4 dots in the top right circle. They represent ‘Smart service via technology’ (Betweenness Centrality: 0.2458), ‘Equal access to services’ (0.2338), ‘Health, Well-being’ (0.18113), followed by ‘resource management’ (0.12168). The number of action initiatives under “resource management” is greater than actions under “health and well-being”, in descending order of significance. Betweenness Centrality also reveals the same results in the survey stage (Table 5.16). This means that a node of ‘health and well-being’ signifies the extent of its influence on the interactions among other nodes in the network. Despite the fact that this analysis pertains to network data rather than actual performance results, it is still assumed that technology has higher significance in Nagpur proactively using technology (information) capital, which creates value over time. Consequently,

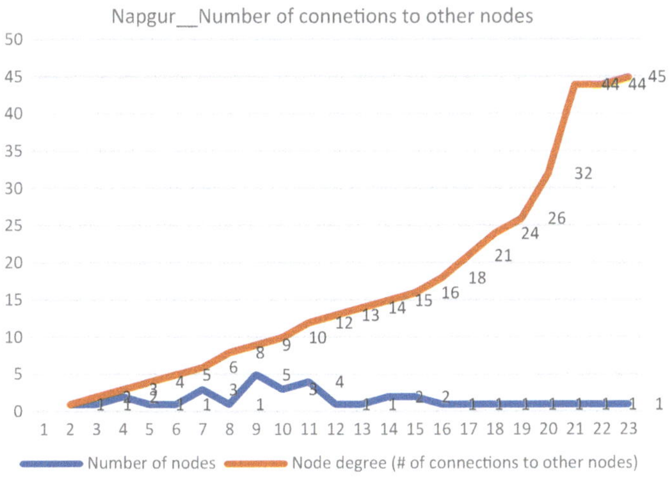

Fig. 5.15 Number of connections to other nodes (Wataya and Shaw 2022)

the findings from in-person interviews and survey results align with the outcomes presented in the proposal stage.

The same applies to Bhubaneshwar, the results of the network analysis can be verified by Nagpur's Ease of Living (EOL) results. The results demonstrated that 'Safety and security', 'City resilience' outweighed the other factors. The scope of the categories used in the EOL did not exactly correspond to 'Resilience' and 'Security' in the classification of key value creation pillars in this study. With that understanding, Nagpur focuses on technology for better service environment, closely corresponding to 'smarter services via technology'. Factors such as 'education; and 'housing and shelter' correspond to 'health, well-being' and 'equal access to services' respectively. In the case of Nagpur, the results of the network analysis are verified. Also the contribution of soft assets can also be validated in terms of the EOL.

Soft assets and resilience

Maharashtra State is prone to various extreme hazards. The State was the first in India to create a Disaster Management Unit after the Latur earthquake in 1993 and then evolving its disaster management approach from a reactive to a proactive strategy. The Maharashtra State Disaster Management Authority (ADMA) is responsible for formulating policies and guidelines for disaster management and developed the state disaster management plan with the goal of creating a safe, disaster-resilient state. ADMA's approach is based on a proactive, technology-driven, multi-disaster strategy that emphasizes a culture of prevention, mitigation, and preparedness, prioritizing the saving of lives and the reduction of property and environmental damage. (SDMP 2015. https://maharashtra.gov.in/Site/upload/WhatsNew/SDMP-2015%20(1).pdf). The District Disaster Management Authority (DDMA) is responsible for disaster management at district level and prepared the District

Table 5.16 Proposal stage and interview stage comparison: Nagpur

Nagpur

	Proposal stage ranking		Number of actions/initiatives	Betweenness centrality (BC)	Ranking (BC)
Total number of actions/initiatives	1	Smarter services via technology	67	0.326	1
	2	Equal access to services	57	0.187	3
	3	Health, Well-being	52	0.272	2
	In-person interview stage ranking		Number of actions/initiatives	Betweenness centrality (BC)	Ranking (BC)
	1	Smarter services via technology	45	0.245	1
	2	Equal access to services	44	0.233	2
	3	Resource management	44	0.121	3
	4	Health, Well-being	32	0.181	4
	5	Resilience	26	0.084	6
	6	Quality communications	24	0.113	5

Source Created by the author based on the survey, Wataya (2023)

Disaster Management Plan, Nagpur District, 2017–2018, enhancing preparedness, regular monitoring and efficient Human Resource to manage disaster response. The plan addresses disaster risk reduction, official training and development, awareness initiatives, human resource management and cross-departmental coordination for incident response and relief planning. (District Disaster Management Authority Nagpur).

The Nagpur Municipal Corporation (NMC) is responsible for provision of basic urban services and handling disaster preparedness and response within the city. In accordance with the Disaster Management (DM) Act of 2005, Nagpur has been designated as the regional headquarters responsible for coordinating disaster management in the city and its vicinity. This involves setting up a disaster management committee and a city control room in coordination with State, district, and police disaster control rooms, creating ward-wise disaster management plans, and establishing an effective communication system for public outreach. NMC published the Disaster Management Plan 2019–2020, covering situation analysis, vulnerability, hazard profile, potential hazards, and institutional arrangements. (ICLEI South Asia, 2021, 2.-Climate-Resilient-City-Action-Plan-Nagpur-Report-Low-Res_compressed.pdf (iclei.org).

Nagpur district is also vulnerable to various extreme hazards. Because of its unique geographical location serious floods occur on a regular basis, including the Vidarbha floods in 2020 with severe flooding and population displacement. The location of the river, the catchment area, and the dams cause the downstream areas of Maharashtra to become severely flooded. With heavy rains the catchment areas of the Wainganga river basin overflow into the many tributaries and dams. When the floods reached the Nagpur and neighboring districts, more than 26,000 people from 4,500 families were evacuated to safer locations. The Central Water Commission (CWC), the central city organization responsible for monitoring dam storage and flood forecasting, is located in Nagpur. The command and control centre (COC) at Nagpur Municipal Corporation (NMC) monitors the storm water, flooding and discharge storm water using CCTV surveillance cameras installed across the city (The Times of India).

The storm water overflows and flooding can be attributed to several factors—blocked drains, clogging, lack of an effective sewage and stormwater network, and the unique geographic characteristics of the area where flooding can rapidly escalate. Given this vulnerable situation, it is appropriate for Nagpur to prioritize improved monitoring systems, accurate weather data, collection and analysis, dam management and flood monitoring to effectively plan, prepare for and prevent these types of disasters (Tables 5.17 and 5.18).

The importance of proper infrastructure, institutional mechanisms and community engagement on daily solid waste management was also raised where soft assets are

Table 5.17 Successful way incorporating soft assets

Nagpur (Question 8–1)	People/ Human capital	Institutional capital	External capital	Technology (Information) capital	Social capital	Total
Resource Management	5	5	1	4	1	16
Accessibility (Mobility)						0
Health Well-being	1	1		2		4
Equal Access to Services	3	3	1	5	1	13
Smarter Services via Technology	2	2	1	4	1	10
Community Engagement	1	1	1			3
Quality Communications	1	1	2	1	1	6
Security	3	3		4	1	11
Resilience	5	5	2	6	2	20
Total	21	21	8	26	7	83

Source Created by the author based on the survey, Wataya (2023)

Table 5.18 Three instances of soft (intangible) assets crucial for enhancing resilience

Nagpur (Question 8–2)	People/ Human capital	Institutional capital	External capital	Technology (Information) capital	Social capital	Total
Resource Management						0
Accessibility (Mobility)						0
Health Well-being						0
Equal Access to Services	2	2				4
Smarter Services via Technology	1				1	2
Community Engagement	3				2	5
Quality Communications	2				2	4
Security						0
Resilience	6	4			2	12
Total	14	6	0	0	7	

Source Created by the author based on the survey, Wataya (2023)

requested to improve resilience. According to ICLE's Climate resilience city action plan-Nagpur (2021), the risk assessment results identified fragile urban systems with water supply, solid waste management, rated as high risks. Embedding soft assets in infrastructure projects is vital, not only for reducing carbon emissions, but also enhancing resilience to disaster and climate risks.

The survey questions asked for a description of successful ways the city incorporated soft assets to build resilience during the Smart City Mission implementation period. The responses listed technology, communications, health, well-being and included security and safety infrastructure with CCTV, social media, public announcement, use of green environment and natural rivers, open spaced, plantations for different adaptation, mitigation and precautionary measures to reassure local communities. Also identified to improve resilience and services were more effective management of storm water networks and establishing thermal power plants to minimize pollution, contamination, and dust pollution. These actions are not only related to 'resilience', but also to 'equal access to services'. This means that embedding soft assets in specific disaster actions influence other soft assets value elements. In this context the Nagpur survey results revealed the following influential nodes: 'smarter services via technology', 'equal access to services', 'quality communications', 'resource management'.

5.3.3 Value Creation Mapping

Identifying cases illustrating value creation

Similar to Bhubaneswar, value creation mapping was also conducted in Nagpur. Identical questions were used for the survey. Responses to questions about the successful examples and incorporation of soft assets capitals and the impact shows a compound, multilayered, and holistic contribution by soft assets.

Table 5.19 shows the classification of the respondents whose answers are summarized in Table 5.20, using the qualitative framework adopted in the network analysis. The results of the classification show that 'Smarter services via technology' and 'Equal access to services' are also significantly influential nodes in this mapping exercise and consistent with analysis from the proposal stage. The results also verify the findings of the connection (network) analysis at the proposal stage. As with the case of Bhubaneswar, the use of soft assets creates benefits and value for the citizens and the city as a whole.

At the proposal analysis stage, there was one influential node not included in the nine value elements, which was 'consensus building'. This is categorized under 'quality communications' and 'community engagement' which also verifies that the influential node identified at the proposal level contributes to value creation.

Table 5.19 Classification of value creation actions: Nagpur

	People/human	Institutional	Tech/information	External	Social	Total
Resource management	3	2		1	1	7
Accessibility (mobility)		1	1			2
Health, Well-being	1	4	1	1	1	8
Equal access to services	1	6	3	1	1	12
Smarter services via technology		7	6			13
Quality communications	2	3		2	2	9
Community engagement	2	2		2	2	8
Security	1	3	1	1	1	7
Resilience						0

Source Created by the author based on the survey, Wataya (2023)

Table 5.20 Some responses from interviewees regarding positive action and its impact

Positive Impact (1)	Exploring their knowledge and working on new challenges and advanced components that will definitely be useful for them in their future career	Crime Detection and Prevention, Monitoring, and Maintenance of Law and Order Situation	These helped a lot in enforcing citizens to obey the traffic rule, resulting in people starting to wear helmets, seat belts, refrain from jumping signals, as they now know CCTV is watching them It has also aided significantly in the detection of crime	Integrating soft aspects (i.e. citizen-centric approaches)	Data collection from relevant stakeholders pertaining to sectoral energy and fuel consumption and evaluation of GHGs following GPC protocol	Will offer low-cost parking space and support better mobility and accessibility in crowded areas	Required effective coordination among different line departments and organizations such as SPV, NMC, Electricity Board, etc.; Will support better and integrated service delivery to communities as the smart roads would not require regular digging (which is the current practice for putting pipelines of different utilities) till at least next 10–20 years	Helping citizens as well as city administration in reducing wastage of electricity, proper garbage collection, short waiting time on traffic signals, etc

(continued)

Table 5.20 (continued)

Positive action (1)	Opportunities for employees/ staff working under the smart city **[RM x Ppl/ HR]**	More than 3600 cameras were installed at 700 junctions, along with multiple ICT- based components connected with 1030 km of optical fiber cable network installed by NSSCDCL **[SMART x IT/ INFO]**	ANPR/RLVD and other analytics-based software, along with related hardware, were installed for crime detection and challan generation **[SMART, SEC x IT/INFO, INSTIT]**	Encouraging more eyes and life on streets and in the city, which not only makes the cities more lively but also provides a sense of security to communities, changes the course of conversation to positive and healthy choices, etc **[SEC, QC x INST, SOCIAL]**	Preparation of the city's GHG emissions inventory baseline **[H-Well b, x IT/INFO, INST]**	Multi-level parking systems in busy market areas **[Access, SMART x IT/ INFO, INTST]**	Integrated smart road with underground utilities (water supply, power cables, sewerage pipes, etc.) **[ACCESS. SMART, EQ x IT'INFO, x INSTI]**	The use of the ICT-enabled components **[SMART, EQ x IT'INFO, x INSTI]**

(continued)

Table 5.20 (continued)

Positive Impact (2)	The citizens under the ULB are performing best practices within the framework of the Smart City initiative	Through monitoring wells that measure different water parameters and level throughout the year, a mechanism has been put in place for real-time remote monitoring, aiding in informed decision-making	Helping the city administration better understand ground- level bottlenecks and gaps	Enabling scientific and informed decision-making	Mapping of wards that are more vulnerable to different climate risks and hazards is conducted based on stakeholder consultation, secondary research, and the use of GIS techniques	Identification of potential areas for GWR is done using GIS techniques, primary surveys and secondary research. The use of vertical electrical sounding techniques helps identify the optimum depth below ground for recharge	This will benefit the local communities that were earlier facing health and environment risks such as fire incidents, smell, leaching, particularly during monsoon	People were engaged in the identification of cycling routes and areas. It has enhanced awareness and acceptance of environmentally friendly cycling infrastructure among the public
Positive action (2)	New, advanced things to serve	Rainwater harvesting **[SMART, x it/ INFO, INST]**	Consultation and feedback with beneficiaries/ user **[QC, CE, BS x IST, HR, EST, SOCIAL]**	Enhanced capacities of ULB staff **[RM x PPL/ HR, INST]**	Climate Risk and Vulnerability Assessment **[H, WELL, RES, SEC, QC, CE x PPL/HR, INST, EXT, SOCIAL]**	Implementation of Rain Water Harvesting (RWH) and ground water recharge (GWR) systems at 2 public schools **[SMART x INST, IT/ INFO]**	Reclamation of land (waste landfills) **[H, WELL-B, BS x INST]**	Cycle for change **[H, WELL-B, CE x INST]**

(continued)

Table 5.20 (continued)

Negative impact (1)	However, not much of local consultations and need assessment were undertaken involving the local communities. As a result, the developed resources are not being utilized optimally, and their impacts remain unknown due to a lack of active data and information sharing	Their effectiveness is not yet established, as there is no result or data sharing on how the findings are further used	A major lacuna is the absence of night mode in the CCTVs, which makes them less effective during nighttime	A major challenge in their utilization is the lack of proper planning and management of the useful content that should be displayed on these screens	It is observed that sensors in many of them are not fully functional, leading to spillage and piling of waste around the dustbins. Possibly, solutions installed in the city should be as per the city's requirements and capacities to effectively operate and utilize them	However, the suggestions and feedback are not yet reflected, and these kiosks are reportedly used by beggars		

(continued)

Table 5.20 (continued)

Negative action (1)	Under the pan city development, there have been some good initiatives undertaken	The environmental sensors are installed which records different environmental parameters	CCTVs installed across the city are supporting constant monitoring of law & order and hazardous situations in the city,	Digital screens have been installed across the city,	Smart bins installed across the city are expected to send alert to the local waste management authorities when they are filled	Kiosks have been installed near the bus stop at various locations of the city		

Source Created by the author based on the survey, Wataya (2023)

The relationship between the five soft asset capitals and the value creation process

As with Bhubaneswar (Figs. 5.7 and 5.8 in this Chapter), these are some examples set out in Table 5.20 and the Nagpur Smart City Mission proposal that demonstrates how soft asset capitals are inter-related in the process of value creation. Figures 5.7 and 5.8 show a comparable pattern of value creation and value non-creation. The key summary information is listed below.

(a) Value created case (positive case)

- Activity/project: Multilevel car parking system (Smart Parking Sub-system)
- Objectives:
  - To accommodate large number of different makes of cars in busy areas
- Inputs:
  - *Soft asset capital*: Technology, information capital, Institutional capital
  - *Value elements:* Equal access to services
- Output:
  - Offering low-cost parking space, supporting better mobility and accessibility in crowded areas
- Outcome & future impact:
  - Improving technology and applications including enabling electric car parking system through a network of sensors, digital display boards at various junctions; information system for available parking places in the vicinity
  - *Soft asset capitals*: Technology, information capital, People/Human capital, Institutional capital; Community engagement, External capital
  - *Anticipated Value elements:* Equal access to services, Smarter services via technology, Community engagement; Resource management

(b) Value not created case (negative case)

- Activity/project: Smart garbage management systems (Smart Swachh city solution)
- Objectives:
  - Smart refuse bins installed across the city with fixed sensors and wireless communication links that remotely send alerts to the local waste management authorities when they are filled and ready to be collected
- Inputs:
  - *Soft asset capital*: Technology, information capital, Institutional capital, People/Human capital,
  - *Value elements:* Equal access to services, Smarter services via technology,

- Output:
  - Failure of appropriately functional sensors leading to spillage and piling of waste around the bins
- Consequences:
  - Wasting investment materials, budget, and existing facility; Lost opportunity of value creation
  - Hygiene issues and public health concerns
- Possible reasons for opportunity failure:
  - *Soft asset capital: weak institutional capital and technical and information capital to set standard requirements of installed equipment, people/ human capacity to operate and utilize the equipment*

Value creation mapping (canvas)

As with Bhubaneswar, using Table 5.20 in the previous section, value creation mapping was created and shown in Fig. 5.16 below. The maps capture only a few of the activities that are being implemented. The development of a more accurate value creation map needs additional information such as timing, classification of different stakeholder involvement and linkages to other initiatives and planning. This additional information could help further expand this value creation map, including assumptions of scaling up, synergetic, and spillover effects. For this purpose, it requires a series of stakeholder consultations to check the progress of follow-up actions, as well as strategy and systems to integrate the results into the planning both at strategies and project levels, thus providing a feedback mechanism. The value creation map below demonstrates the effects of incorporating soft assets taken from the survey results. Major common findings from this visual mapping (canvas) are listed in the Bhubaneswar value creation mapping section.

5.3.4 Consultation, Quality Communications, Community Engagement, and the "Learning and Share' Process

Regarding consultations and communication, the same questions used in the Bhubaneswar survey were also used for the Nagpur city administrators, SPV staff, and other stakeholders. Nagpur also places importance on consultations, but there are different findings from Bhubaneswar in terms of the approach and focus of consultations as well as the institutional arrangements regarding learning and share activities.

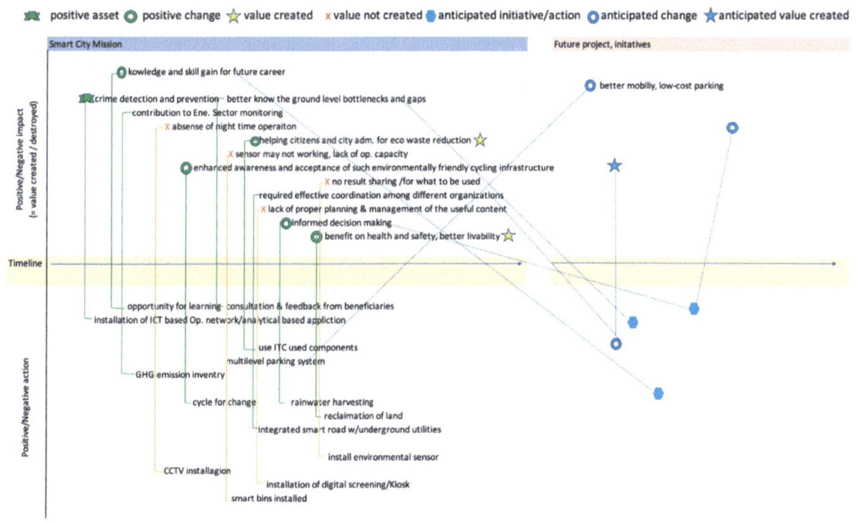

Fig. 5.16 Value creation mapping (Nagpur) (*Source* Created by the author based on the survey, Wataya 2023)

5.3.4.1 Consultation Process, Quality Communications, and Community Engagement

Nagpur city has also conducted extensive consultations through smart city related events to maximize citizen engagement. The consultations were conducted by a media campaign to collect views and suggestions from citizens as well as focus group discussions on various city issues and challenges. Workshops, a variety of meetings and citizen engagement programs, in parallel with various strategic stakeholder consultation workshops were undertaken to gather inputs and suggestions for integrating smart city features into city operations to enhance the overall citizen experience in the city. Nagpur city has engaged various stakeholders to reach a consensus on the vision and ideation process as part of the preparation of its Smart City Mission Proposal. (https://blog.mygov.in/nagpur-municipal-corporation-citizen-engagement/) Some of the activities were captured from the survey responses below.

(a) **Entry Points**

From the survey responses, Nagpur also conducts various levels of consultations to promote inclusiveness and awareness of its smart city initiatives. Integrating community and different stakeholder's needs are crucial to Nagpur's planning for smart city development. Some responses indicate that the most intensive consultations take place at the proposal stage, but comprised only briefings in later stages. With only limited responses on this issue, it is still clear that there are expectation gaps on the roles of consultations between different respondents i.e. local officials and academia, NGOs. Nevertheless, there are explicit responses regarding institutionalized consultation and feedback, but not a mechanism to address grievances (Table 5.21).

(b) **Effectiveness of Consultations**

Major points contained in the responses are awareness raising, understanding of project impact, consensus building for new plans and an institutionalized grievances mechanism for citizens. Consultations are more common in the project planning stage to seek early citizen buy in and support for local projects and activities. However, there are no more detailed aspects of inclusiveness in the limited survey responses (Table 5.22).

(c) **Successes and Challenges**

In Nagpur city, there were several factors identified from the survey results, including the use of technology applications, which were found to be more important for Nagpur than Bhubaneswar. Some key points from the survey are also listed in Table 5.23 below. Again, the survey results did not capture every aspect, but difficult cases were identified such as relocation and compensation issues. This strongly correlates with the soft asset capital factors of 'quality communications' and 'community engagement' and remain challenges in Nagpur.

- Community engagement is inadequate, without detail or institutionalization.
- A few community-based surveys were undertaken, but not sufficient to gauge the views of the wider (local) communities.
- The Smart City Planning in the area proposes Area Based Development (ABD) partly involving road infrastructure, but would result in the removal of around 400 people. It appears that the affected people were inadequately informed or engaged with the effect of the planned development.
- There also appears to be an inadequate relief and rehabilitation policy for the ABD area, in particular, information gaps on criteria used for affected people allotted alternate housing and entitlements.

5.3.4.2 The "Learn and Share" Process

(a) **Learn and share process in operation**

Many survey responses confirmed that a prime focus of city authorities is delivering physical or hard infrastructure which is tangible and easy to describe. Responses also reiterated that soft assets such as human resources and skills development are accomplished mainly through training programs. The biggest differences from the survey responses of Bhubaneswar is there are no separate mechanisms for learning and sharing like BUKC in Bhubaneswar, or at least there was no feedback regarding

Table 5.21 Process of consultation (implementation)

Purpose	Description	Entry point
Demonstration of successful cases	Include the best practices from them in our Smart City	All parts of the citizens
Enhance communications and engagement of stakeholders	Citizen Advisory Forum supporting the consultation, communication and engagement with community and other stakeholders. The Forum meets every 3 months in this regard	Members of NMC, local representatives with sectoral expertise (for example, education, commercial operations, NGOs, etc.)
Policy formulation, debriefing	Public consultations are mostly done at policy formulation level At implementation level, there is no institutionalized mechanism of seeking consultations and feedback except for a Grievance Redressal Mechanism	Draft versions are open to diverse stakeholders for comments and feedback
Project design	Primary surveys through web and physical to assess citizens' perspectives and expectations to design and implement projects At the project planning phase, the consultation promotes participatory governance	Academia, local CSOs, NGOs, etc
Livability, KPI	Livability Index, Cycle for Change, Streets for People, etc	Extensive stakeholder consultations, field surveys, citizens dialogues, etc
Benchmarking and assessment	Research/survey to benchmark and assess the existing urban services zone-wise	Under NMC coordinated
Consensus building and seeking commitment	Undertake (or at least commit to undertake) stakeholder/community consultations during proposal stage. But similar consultations are hardly seen during implementation stage	
Gaining support re. smart city planning and activity	Support consultations and seek advices/feedback from different stakeholders on different planning and activities under the Smart City Mission. The forum is expected to meet quarterly, however, sometimes, the meetings are held once in six months	Government representatives and political representatives such as Member of Legislative Assembly/Council (MLA/MLC), Mayor, etc., of the respective area. The non-governmental members of the forum include representative of NGO, Residential Welfare Association (RWA), and that of local hotel associations

(continued)

Table 5.21 (continued)

Purpose	Description	Entry point
	Consultations are more visible during the proposal stage, the organizations bidding also commit to undertaking consultations with local communities and other stakeholders, etc However, the same visibility is not much observed during the implementation stage	
Awareness raising	Supports the programs and activities of the Nagpur Municipal Corporation, where they are regularly invited and engaged with in awareness programs, workshops, etc	AIILSG

Source Created by the author based on the survey, Wataya (2023)

Table 5.22 Effectiveness of consultations

Effectiveness	* Yes, definitely, these consultation and feedbacks are effective. Citizens helps to implement the project on ground as they were aware of the output and outcomes of these projects
	* Feedback received helps identify the local needs. For example, in the initiative Cycle for Change, the consultations and feedback from the Forum helped identify suitable locations for cycle stands to suit the needs of the local communities
	* Grievances received are used to address identified gap in the respective initiatives. For example, based on grievances received recently, it came to the light that in the name of trimming of old/broken branches/trees during pre-monsoon season, cases of felling of trees were observed in a few areas to fulfil personal motives. Accordingly, the relevant directive by the City Administration was modified adequately to curb such practices
	* They are effective. For instance, local cyclists were consulted extensively to design and implement a pilot project of a dedicated cycle track within the city's jurisdiction
	* However, it is felt that rather than consultations, the members are informed of the planned activities of the Smart City
	* It is also felt that suggestions provided are not adequately incorporated in the planned activities
	* However, there were hardly any similar engagements with Smart City Mission, which involved AIILSG

Source Created by the author based on the survey, Wataya (2023)

such mechanisms except the functions provided by the Command and Control Centre. Some documentation on lessons learnt for future projects and initiatives is produced at the project level. However, mechanisms to prolong these approaches could not be identified from the survey results (Table 5.24).

Table 5.23 Successes and challenges

Factor	Actions
Technology, applications	College Students, parents, government organizations, citizens, etc. were engaged, and individuals were communicated with to achieve success
	A secure and safety infrastructure with CCTV, green environment, and good disasters management facilities available can contribute to get success
	Citizen and communities are well connected through: (i) wide information sharing: (ii) active communication by Nagpur Municipal Corporation using social and other forms of media;
	CPS through web-based forms, easily shared using social media tools: (i) use of local outreach channels such as radios, TV channels, etc.; (ii) outreach through local newspapers
Reaching out broader targets	Reaching out (in person) to stakeholders such as schools
City Level Advisory Forum	The forum such City Level Advisory Forum (consisting of local politicians, students, social and environmental activists, professors, etc.) might have been helpful in enhancing community engagements

Source Created by the author based on the survey, Wataya (2023)

Table 5.24 "Learn and Share" process in operation (in practice)

Level	Entry. Process
Central	Respective divisions share their data and information on the designated portals of GoI Learning and sharing are also supported by different initiatives of GoI, including the publication of good practices, conferences, etc
	Learning and sharing are supported by different Advisory Notes provided by GoI, along with a unified GoI's portal that facilitates cross-sectoral updates and learning
National, SPV	There is no separate mechanism of learning and sharing at the SPV level
	Requirement, availability of ICT components to fulfill the citizens' requirement are submitted as per the survey
	The role of the department is mainly in integration and implementation of the ideas
Project level	* Technical assessments and pilot projects were undertaken based on stakeholder consultations, thereby following the city's priorities * Lessons learnt were documented and shared with relevant city departments in the form of trainings, fact sheets, case studies, etc * Project outcomes and best practices were published on official websites and national institutions and think tanks for better outreach
Citizen,	* Citizens are involved to gather their reviews * Consultants were hired to perform the survey and process the requirements

Source Created by the author based on the survey, Wataya (2023)

(b) Challenges in establishing the "Learn and share" process

From the limited survey results, challenges are not always clearly explained, but some cases indicate that there are some initiatives to facilitate learning and sharing at institutional, and project levels. Using the SMARTNET portal, existing knowledge is able to be disseminated. The effectiveness of the "Learn and Share" process with or without national or city level functions requires not a fragmented, but systematic approach to accumulating and nurturing knowledge, case examples, experience and other key information (Table 5.25).

(c) Evaluation of the vision, goals, activities and monitoring soft assets

The limits of the survey responses made more difficult the accurate assessment of the vision, goals, activities of smart city development, taking into account the contributions or value creation of soft assets. However, monitoring and evaluation of the progress of projects in India's Smart City Mission is reported using India's Management Information System (MIS). Also the Ministry responsible for Smart Cities Missions has set up monitoring and evaluation mechanisms to monitor and track fund flows, project planning and progress of execution. The results are available and cities are ranked along with recommendations for further improvement.

Like Bhubaneswar, respondents from Nagpur also referred to the Livability Index, intended to complement the consideration of soft assets. The Ministry monitors programs on a monthly, quarterly, and annual basis depending on the project involved. For instance, the Livability Index and Climate Smart Cities Assessment Framework are used to assess different indicators of performance with appropriate rankings. Recommendations were made to each city and capacity building workshops were organized to provide further assistance. Soft assets are also monitored and evaluated including citizens engagement, building community relationships and working with NGO's, stakeholders and monitoring physical, financial and fund utilization progress.

Internally, there are various mechanisms such as regular internal meetings headed by the CEO, SPV, and quarterly City Level Advisory Forum meetings with Board Members. These mechanisms can also monitor and evaluate, which is examined in Chap. 4. From the survey results, there are project-based monitoring reports, but the

Table 5.25 Cases and challenges of the "learn and share"

Good case	* To complete the incentives provided by the organization, Smart City Mission takes on various challenges
	* Cities must effectively leverage the expertise and assistance provided by external institutions to undertake climate actions
	* Conduct workshops within cities of similar demographics to further facilitate learn and share process
	* Ensure the effective and focused dissemination of existing knowledge products, such as SMARTNET portal and the outcomes of programs such as Climate Smart Cities Assessment Framework, considering the limitations, interests, and priorities of cities

Source Created by the author based on the survey, Wataya (2023)

overall Smart City Mission implementation monitoring and evaluation reports were not found, published, or identified by survey respondents.

References

ICLEI South Asia (2021) Climate resilient city action plan—Nagpur, https://southasia.iclei.org/wp-content/uploads/2022/04/2.-Climate-Resilient-City-Action-Plan-Nagpur-Report-Low-Res_compressed.pdf. Accessed 8 October 2023

Kawyitri N, Shekhar A (2021) Assessing vulnerability and capacity of Bhubaneswar as a progressive smart-city: An empirical case study of Fani cyclone impact on the city. Int J Disaster Risk Reduction 56(1):101986. https://www.sciencedirect.com/science/article/pii/S2212420920314886

Ministry of Housing and Urban Affairs Government of India, (2021b) Making a city smart: learnings from the smart cities mission a workbook for decision makers. https://smartnet.niua.org/sites/default/files/resources/making_a_city_smart_mar2021.pdf

Ministry of Housing and Urban Affairs, Government of India (2021c), Ease of Living Index 2020: Assessment Report https://livabilitystore175634-prod.s3.amazonaws.com/public/docs/Ease_of_Living_Report.pdf

Ministry of Housing and Urban Affairs, Government of India (2021d) Municipal performance index 2020: Assessment report, https://livabilitystore175634-prod.s3.amazonaws.com/public/docs/MPI_Report.pdf https://smartcities.gov.in/Ease_of_Living_Index_and_Municipal_Performance_Index

Pagano and Neubert (2015) Building resilient SCs: mapping and measuring key value drivers through a multi-perspective and multi-stakeholder value creation framework based on intangible assets, conference paper, Part of the IIP advances in information and communication technology book series (IFIPAICT, volume 463)

Wataya E (2023) Integrating soft assets in smart and resilient city Development in India. Unpublished PhD thesis, Keio University, Japan

Wataya E, Shaw R (2022) Soft assets consideration in smart and resilient city development. Defining and debating on smart cities or smart buildings, smart cities 5(1):108–130. https://www.mdpi.com/2624-6511/5/1/7 https://doi.org/10.3390/smartcities5010007

Chapter 6
Discussion

6.1 Key Findings from the Two Indian Cities Case Studies

This research started with an examination of the growing global trend of countries building smart, resilient cities as a solution for sustainable development. In particular, developing countries have major disadvantages compared to developed countries building an enabling environment for smart, resilient cities. These include access to finance due to budget constraints, lack of investment in basic infrastructure and suboptimal human and institutional capacity. There are many areas where soft assets can play a key role to establish an enabling environment and creating value to make cities smarter and more resilient. Among these are robust technology, infrastructure readiness, governance, regulatory safeguards for smart cities, skilled human capital, inclusivity, and citizen participation. However, as several previous research findings have indicated, the unique and complicated features of soft assets can inhibit proactive support and investment. To better understand how soft assets can be used in projects, five soft asset capitals and nine value elements are selected to develop a qualitative framework to identify the use of soft assets.

This qualitative framework (Table 3.8, Chap. 3) was developed from (i) a public organization's 'three higher objectives' (Fig. 3.4, Chap. 3) namely 'cost of providing services', 'value/benefit of services'/ and 'support of legitimizing authority'; and (ii) some existing key pillars of smart cities derived from diverse classifications and indicators. The nine soft asset value elements used in the framework are: 'resource management', 'accessibility (mobility)', 'health well-being', 'equal access to services', 'smarter services via technology', 'community engagement', 'quality communications', 'security', and 'resilience', serve as a foundation to identifying where the five soft asset capitals, i.e., 'people/human', 'institutional', 'external', 'technology (information)', and 'social'.

To examine the degree and extent of soft asset incorporation, the Indian cities of Bhubaneswar and Nagpur were selected as relevant case studies. Each city varies in

E. Wataya and R. Shaw, *Soft Assets in Smart and Resilient City Development*, Disaster Risk Reduction, https://doi.org/10.1007/978-981-97-3373-6_6

size and geographical features, with different smart city visions and strategies, but both represent some of the best smart city strategies used within India.

Based on network analysis and a value creation mapping exercise, there is generally a link between influential nodes and actual value or impact creation. The major influential nodes also correspond to the result of the EOL ('Ease of Living index') score where levels of soft asset consideration contribute to an improvement of livability. The aspect of resilience is also validated by the different approaches and value elements supported by each city. Also three straightforward and simplified displays of value creation mapping and results were used from an analysis of individual cases where value was 'created'; and 'not created'. Significant influential nodes can be validated with an analysis of different approaches to disaster prevention and risk management, the use of consultation processes, the level of the EOL indicators and value creation mapping results.

Based on the findings in the previous chapters, this chapter discusses: (i) the results from the two Indian city case studies, including a comparison between the planning and implementation stages, linking soft assets incorporation to Government issued indexes—the EOL and the MPI, and the various challenges in incorporating soft assets; (ii) a review of value creation mapping and the "Blackbox" concept; (iii) challenges in smart, resilient city development, including the importance of central coordination mechanisms; and (iv) the optimal timing of the most effective approaches to incorporate soft assets in smart, resilient city development.

This chapter also discusses the importance of assessing "comprehensiveness" within the value creation process, a holistic perspective that considers factors such as the passage of time and variations in the layers of the city. The results demonstrate that the process of soft asset value creation evolves organically. Therefore, the process of urban soft assets integration should be analyzed both at the project level specifically and the citywide level as a whole to determine the effectiveness of resilience policies and long-term sustainability. Finally, the survey findings and analysis argues for a need for a "co-value creation" model.

Comparison between planning and implementation stages

In each city, the influential nodes or value creation areas that were identified in the proposal stage were also identified in the implementation stage, with some differences. Although the identified nodes of both cities were strongly aligned overall with smart city objectives, the two stages (proposal/planning and implementation stages) in Nagpur were more consistent than in Bhubaneswar.

This research acknowledges the inherent limitations in covering all aspects of soft assets in a comprehensive way. Recognizing the importance of capturing soft assets within various contexts, connecting them holistically and avoiding individual analysis for each soft asset, the results are derived from a comparison of a limited set of responses. As demonstrated by Bhubaneshwar's responses, the differences in the results, particularly the emphasis on the soft aspects during the implementation stage in the two cities, likely contributed to the variations.

Overall, the range of responses demonstrates a snapshot of how soft assets have been incorporated in each city. The different results from the planning and implementation stages can be attributed to several factors.

The responses are to some extent subjective. Different influential nodes are referred to in the proposal stage applied to soft assets. Responses about soft assets incorporation assessment depended on the city administrators' and other key stakeholders' professional perspectives, experiences, observations, and insights at different decision making levels. Despite the different roles of respondents in the Smart City Mission, some of the responses were more detailed than others. The survey also polled the experience of city administrators about specific items, including the use of 'consultations', the 'learn and share' process or achieving 'resilience'. Thus, different insights, experiences, and challenges were identified in each city about how soft assets are perceived as a key factor in urban planning and development. The results of the analysis and comparison from each city showed clear differences in approach in integrating soft assets.

From the survey, one of the biggest challenges for respondents was to understand the terminologies used in this research and the questions often needed to be modified and explained for easier comprehension. It is clear that 'people/human capitals' and 'institutional capital', including system and processes, are important enablers for respondents to make projects operational and these are essential soft asset capital elements. The use of technologies to establish better relations between city authorities and communities, diverse communication channels, citizen and community engagement, were also identified as key elements for a smart, resilient city.

The responses often did not indicate a utilization of soft assets, anticipated outcomes or their impact on project or activities. Soft assets were incorporated at different levels of policy, project planning, and implementation. As discussed in Chaps. 1 and 3, it is difficult to attribute specific inputs to outcomes or impact. Nevertheless, some responses provided useful examples of anticipated impacts that will be discussed further in the value creation mapping section. It was also found that both cities identified areas requiring the utilization of soft assets recognizing their importance in raising awareness for sustainable development.

Refining the survey questions to apply to the more specific project level was challenging. On the one hand, examining the details of the physical implementation process in a city has been already monitored and examined. On the other hand, the use of soft assets has already been factored into pre-existing decisions and it is therefore more difficult to identify specific factors which may have influenced city administrators. Hence, the differences in individual comprehension and interpretation of questions reflected in the survey outcomes were dependent on their duties and work program.

The Impact of Utilizing Soft Assets

Overall, this research demonstrated that the cities most frequently employed the following soft asset value elements: the use of 'smarter services via technology', 'quality communications', 'equal access to services', and 'community engagements'. In addition, building 'resilience' is also significant. The geographical location and specific disaster events experienced by the cities also affected the focus of the disaster prevention and response and influenced how they employed soft assets to enhance resilience.

This research found that different influential nodes were important and demonstrated that the incorporation of soft assets was aligned with each city's specific vision, objectives, and strategy to create value. In Nagpur, for example, 'human resources', 'institutional capital', and 'external capital' in 'quality communications' were used to plan and operate more efficient waste management systems. The synergetic, spillover, and scaling-up effects of these embedded soft assets in the value creation process also impacted other areas of city management as well.

The greater number of preferences in the responses for a specific element does not necessarily mean it is more important than any of the other nine soft asset value elements with lower numbers of responses. These elements are not on the same level—some comprise cross cutting themes (i.e. 'resilience'), some are more fundamental than others (i.e. 'resource management'), and some involve a process of relationships (i.e.'community engagement'). Also respondents had different perspectives and priorities within the nine soft asset value elements due to their own situations.

In the value creation process, identifying the influential nodes is important to understand where soft assets incorporation is more concentrated than others and managed strategically. However, the quality and values are context specific and the level of scalability and ultimate impacts are unpredictable. Therefore the results of the comparative analysis present only a snapshot of the various city approaches.

Linking soft assets utilization to Government issued indexes—EOL and MPI

Using soft assets is one of the key contributing factors to improving government performance and this can be measured by the Ease for Living Index (EOL) and the Municipal Performance Index (MPI). These indexes used by India provide a reliable measurement of soft assets used to build capacity and create value over time. They are important for the evaluation of India's 111 cities to determine the gaps in implementing urban development programs and to fine tune strategies to achieve their goals (MoHUA, MPI 2021). The two indexes cover key 'pillars' and 'a set of five vertical' categories which encompass core functions i.e. for EOL: quality of life, economic ability, sustainability; for MPI: governance, planning, technology, finance, and service delivery.

The two Indian case study cities have already demonstrated solid performance outcomes using the EOL and MPI evaluations. From the EOL ranking—Bhubaneswar was ranked 2nd among 62 cities for cities of less than a million population and Nagpur was ranked 25th among 49 cities with populations over a million. Using MPI, each city has different areas or enabling factors that score higher than the national average—Bhubaneswar ranks high in 'planning' and 'technology' while Nagpur ranks high in 'technology' and 'governance'. The top three city scores are categorized under sustainability (e.g. city resilience) and quality of life (e.g. housing and shelter, safety and security, education).

This data-based evidence demonstrates that both cities utilize some enabling factors (represented by MPI indicators) to deliver effective, valued services to their

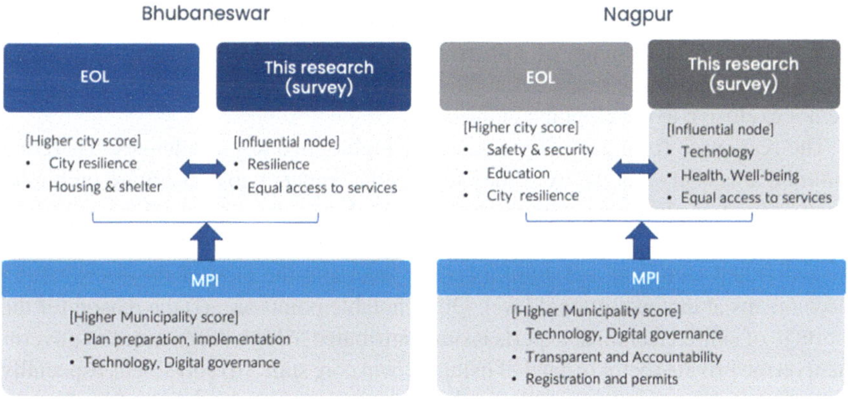

Fig. 6.1 Verifying influential nodes with EOL and MPI (*Source* Author's original work, drawing upon information from MoHUA, Government of India, Ease of Living 2020. Municipal Performance Index 2020)

citizens (represented in the Citizen's Perception Index). Taking into consideration the EOL pillars and the MPI's set of five verticals components, soft assets prove to be significant contributing factors to achieving each city's EOL and MPI results. A broader linkage can also be identified between the EOL score in each city and the influential node identified from the survey results.

Figure 6.1 shows that higher EOL score items correspond to the influential nodes mentioned in Chap. 5. The EOL scores encompass many other factors that influence the score. However, it also indicates that the level of soft assets consideration is one of the main factors contributing to the score. As an enabling factor, the plan preparation and implementation capacity of Bhubaneswar (in this research, this equates with 'institutional' capital) is used in several areas to develop smart, resilient city planning and implementation. Including the MPI results on 'Technology' such as strong digital governance and some 'services' sectors, Bhubaneswar's outcomes were validated by the highest results it received in the Citizen's Perception Index.

A stronger technology focus is employed on Nagpur's smart city proposal and implementation stages. The obvious development impact and values are created by delivering effective services using innovative technology. Other aspects that make Nagpur unique is the existence of a strong service sector resulting in 'transparency and accountability', 'education', 'registration and permits'. This was also supported by Nagpur's performance score in 'governance' and 'service' which is higher than the national average. These enabling factors from the MPI also correspond to the survey results.

Challenges in using soft assets

There are fundamental challenges that interviewees shared in the survey, which relate to institutional systems, enabling factors including human issues and the effect of bureaucratic silos. The results correspond to the challenges faced by developing countries examined in Chap. 1. More importantly, these challenges are due to a combination of the features of soft assets and public sector operations, including where strong political influence is at play. It is essential to note that these challenges are not exclusive to developing countries.

The responses from each city case study highlighted key challenges. There are common areas between the two cities such as the tendency to focus more on implementing hard or physical assets (infrastructure) due to the easy visibility or lack of political intervention and support. Also highlighted were the limited capacity at both the individual technical and administrative level and the lack of the coordinating mechanisms at the institutional level. Other notable points are: (i) the reason for the shortage of skilled technical experts is often attributed to high dependence of government on the private sector instead of using its own core staff; (ii) soft assets, especially human capital aspects, are usually implicitly considered in planning and implementation; and (iii) ad-hoc approaches of considering soft assets includes disincentives for investment and lack of political will.

Institutional challenges such as changing the public sector is not easy and implementing reforms takes time. Respondents however, felt these issues should be dealt with at earlier stages to establish an enabling environment. It is therefore important to understand what type of soft assets are incorporated, when and where, as this will be a starting point to accomplish desired outcomes and impact on projects and initiatives. There is a need to accurately measure the effectiveness of soft assets outcomes. Measurement of individual cases requires identifying which capital will be measured, and then establishing a benchmark to compare the results and outcome levels. In this regard, the network analysis and value creation mapping tool provides a useful basis to measure soft asset utilization that can effectively demonstrate tangible results. Influential node classification can also help determine specific soft asset value elements and also be one of the possible entry points to develop a process to mitigate anticipated risks.

6.2 Mapping—Inputs, Outputs, and Outcomes

Value creation process

As shown in the earlier discussion of value creation mapping in Chap. 5, many unpredictable factors can minimize the value of soft assets. To achieve desired outcomes, therefore, cities need to consider a variety of innovative approaches for managing such risks. The process may take time depending on the individual city context, capacity and presence of enabling factors. Chapter 3 discusses incorporating soft assets which foster better relations between cities and various stakeholders by achieving three higher objectives of city administrations. City administrators need to identify and prioritize the development and utilization of soft assets, and take risks to benefit from the value creation process.

Value creation mapping provides an immediate visualization or illustration of survey results and provides a valuable mid-to-long term tool to understand the value creation process. This template was used by Pagano and Neubert (2015) as a useful tool for value chain analysis and was used effectively in this research. This mapping tool will also be useful in the future for policy makers and stakeholders to understand 'what' soft assets are incorporated and 'when' and 'how' impacts can be made. The tool is also invaluable in two areas: (i) establishing a process to measure soft assets value creation; and (ii) understanding the importance of both the value creation process at an individual project level, as well as an appraisal of value creation impacts within the city as a whole. Thus the process of using soft assets to create city value is not cumulative, but organic and evolves despite uncertain factors. As explained in Chap. 5, the process constantly evolves and the consideration and adoption of soft assets in the earlier stages of planning and implementation will have the greatest impact on creating future value.

On the first point, the best way to assess soft assets value is to identify those embedded in the specific context to be measured, whether in infrastructure initiatives, action or projects. Next is to understand the context, then set a benchmark followed by a timeline to compare any performance improvement, or changes. This mapping exercise can help identify the entry points and anticipated approaches, enabling the evaluation of outcomes, including the consideration of soft aspects.

On the second point, it is difficult to draw this conclusion from the literature review and survey results due to the level of complexity involved. However, Fig. 6.2 below tries to incorporate the value mapping results in the 'Blackbox' presented in Chap. 1 to demonstrate in a simplified way what may occur. The 'Blackbox' represents the importance of the time factor in the equation with input and output flows. The initial incorporation of soft assets can also represent 'inputs' portrayed in the figure below, left section. The results 'outputs' and 'outcomes' can be identified in the upper left section within the time boundaries of the Smart Mission program. The 'outputs', 'outcomes' or 'impact' identifies future 'outcomes' or 'impact' below in both upper right sections. Thus value creation progresses with time.

The Blackbox illustrates a simplified linear linkage of value creation mapping. **In reality, connecting one set of soft assets with others is more intricate due to synergetic, spill-over, and scaling-up factors, making it difficult to capture in full**. As mentioned, it helps to understand the actual and anticipated flow of individual actions involved, but the value creation at city level is not simply an aggregate of individual actions. This finding is supported by the literature review about the definition of soft assets, their synergetic, spillover, and scale-up characteristics and the unpredicted risks in specific contexts at different levels, and the notion of 'wholeness' discussed later in this chapter. In addition, the value creation process needs to be reinforced by internal growth and learning mechanisms. This will be discussed in the later part of this chapter.

For example in Bhubaneswar, 'self-defense and leadership training' ('people/human capital', 'social capital') resulted in 'confidence building and developing leadership' (value creation). In this case, input and outcomes (or impact/value) would change if unpredictable events occurred. The value created in this case can be a

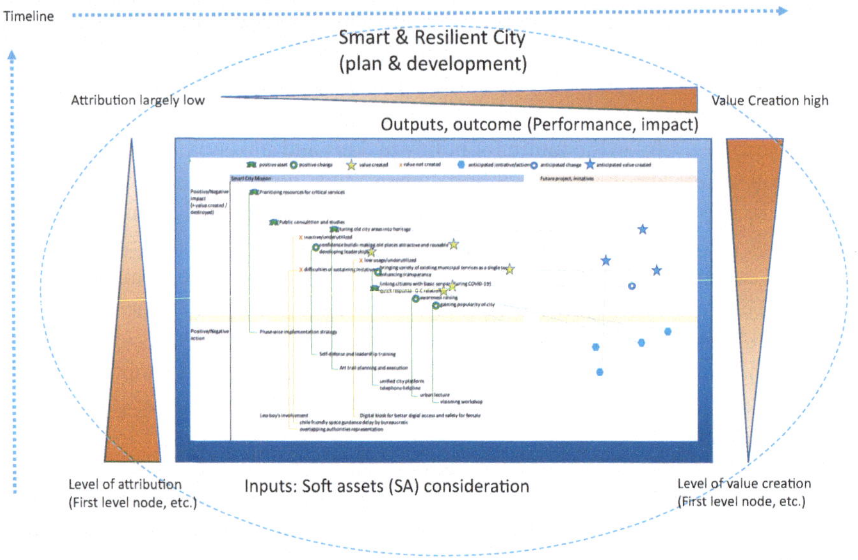

Fig. 6.2 Example of the contents of the 'Blackbox' (*Source* Wataya 2023)

foundation for new initiatives planned in the future. Future action will also be affected by unpredicted changes or risks and may create new synergetic effects with other soft assets in different contexts. This example demonstrates that soft asset value creation is a complex, dynamic and organic process. Important to soft asset value creation is that different outcomes occur at both project and city-wide level. This point will be discussed in more detail later in this chapter.

Along with positive cases with successful results, there were several negative results in two value creation mapping exercises shown in Fig. 5.8, Fig. 5.9 (Bhubaneswar) and Fig. 5.16 (Nagpur) in Chap. 5, presented as 'value not created'. Compared to the original plan, the potential positive results of these actions were not achieved and were categorized as 'value not created', 'failure', or an 'opportunity lost'. It is important to address these failed cases. 'Values are not created' is not necessarily a classification of permanent failure, as they can provide guidance on how to achieve a positive impact if the causes of the negative results are analyzed and utilized appropriately to correct and strengthen using soft assets to create an enabling environment. This requires higher levels of monitoring and evaluation as well as facilitating the 'Learn and Share' process examined in Chap. 5. In this regard, analyzing negative factors can be useful to determine how to create a better enabling environment in future.

6.3 Soft Assets and Knowledge Assets

The characteristics of soft assets as knowledge assets were examined in Chaps. 1 and 3. Also the survey results, showing that knowledge assets are considered critical soft assets, underscores the importance of experiences, cases, and lessons learned for better operational and institutional sustainability. Based solely on the findings of this survey, both cities made efforts at different levels to capture and disseminate knowledge about best practices. The concept of describing soft assets as knowledge assets is included in the five soft asset capital components, in particular, 'people/human capital' and 'institutional capital'. The survey data found institutional mechanisms such as establishing a central city coordinating function, building strong capacity for retaining knowledge and developing sound practices for strengthening institutional capacity, play a positive role in successful knowledge management. This includes collection of information, curation, dissemination, and archiving, all of which have been adopted by Bhubaneswar.

Learning and sharing activities usually take place through the MoHUA portal on data sharing which provides considerable value to each city, although there is a noticeable tendency towards data quantity over quality Therefore the role of the Bhubaneswar Urban Knowledge Center (BUKC) added a considerable value in this area. The BUKC embraces learning and knowledge sharing as a core function by providing central coordination, knowledge and information sharing among city authorities and consultants. The BUKC streamlines the activities across different urban departments and local agencies associated with city operational planning and development, thus avoiding duplication to ensure better resource and knowledge management. They also undertake research on evolving trends and best practices, as well as analyzing cases of Bhubaneswar's successes and failures in the areas of finance, human resources, and administration. However, there are some areas beyond their control. For example, potential political influence and change of city administrators present challenges to the BUKC to consistently support knowledge capture and information development to strengthen institutional capacity. Effectively documenting key features of different planning and implementation practices also helps to develop future administrative and institutional memory ensuring continuity despite management changes.

Furthermore, the data sharing illustrates how to properly disclose data and issue reliable information. Some responses in both cities raised the importance of utilizing the large volumes of data and information collected from different sources, tools, and technologies such as CCTV and RFID. The benefits from 'smart cities' with technology, including technology and its outputs (data), and enabling factors (human, system, resources, etc.) combine well together to produce results. Knowledge as soft assets can provide positive benefits if properly developed and supported in each local city context.

Data accuracy and verification is also important. Data alone is information not 'knowledge'. Also knowledge alone has no value unless verified and integrated into the value creation process. Even though there are large sets of data and information available, especially on environmental issues, it is often inaccurate and unreliable which is a barrier to effective and informed planning and implementation. Thus the information is unable to be used to maximum effect by city administrators. Only validated knowledge should be captured, retained and managed which is a key factor for strategic integration into the overall value creation process. Utilizing accurate data and validated information effectively is critical.

Accurate data and verified knowledge is also covered by the five soft asset capital components especially, 'people/human capital', 'institutional capital', and 'technology/information capital'. 'External capital' also connects to leverage specialized, expert groups or organizations to verify data. The prioritization of soft assets incorporation will be different depending on how data accuracy is addressed and the verification system used. Thus the value creation mapping process can be a useful tool to visualize the effects of the immediate value creation process and flows, but also to show long-term outcomes to further strategically prioritize and nurturing the soft dimension for sustainable development.

Bhubaneswar's centralized knowledge management function has created a system to initiate new, innovative technological approaches to benefit smart, resilient city development while it adapts to unpredictable changes and risks in the value creation process. Establishing a centralized coordination mechanism, described in the next section, provides significant value to an evolving smart, resilient city.

6.4 Challenges in Smart, Resilient City Development

The benefit of a central coordination mechanism is not a new or unique to smart, resilient cities. A central coordination mechanism operates to avoid overlapping efforts within the city administration by breaking up bureaucratic silos across key agencies. This applies not only at city level, but also for example, within the sub-sector of disaster risk management, where regional multi-stakeholder cooperation is required. Here again the value of soft assets comes into play.

As reviewed in Chap. 2, developing countries have major disadvantages compared to developed countries due to fragmented governance, environmental concerns, lack of technological knowledge, financial resources and capacity and skills among citizens. Soft assets integration is vital to create an enabling framework to optimize the benefit of smart city development where technology is fully utilized to improve quality of the lives of its citizens. However, major implementation challenges in most developing countries are technology-related infrastructure readiness, governance frameworks and regulatory safeguards for smart cities, skilled human capital, policies of inclusivity and citizen participation. (Tan and Taeihagh 2020).

The survey results in this research support the following findings:

First, there is often a lack of national and political support in cities for proactive investment in planning and implementing soft assets. The explanation is that the major focus of the government is primarily on physical and hard infrastructure and soft assets, such as staff capacity and human resources are usually considered as inherently embedded in the planning and implementation process.

On human capital, skills development is created mainly through training programs or on an ad-hoc basis, rather than mid to long term planning on the ground. It is important to rely on the private sector, as well as using internal skilled staff available to manage workload due to additional requests from multiple departments. This builds ownership, establishes quality control and sustainability within the city administration itself.

On technology use, both citizens and city staff need to rely on technology to increase service delivery and effectiveness, but often staff are reluctant to change the status quo such as introducing new technology. Also regarding knowledge sharing, there is often a disconnect between the equipment installer (government, other stakeholders) and citizens regarding the benefits of using technology. Another major challenge in developing countries is that technology issues, including basic infrastructure for technology use and its enabling environment, can slow the speed and effectiveness of smart city development.

On institutional arrangements, institutional mechanisms for horizontal coordination among different department and agencies are critical for efficient resource utilization and functioning of major projects like the Smart City Mission. It is essential that cities in developing countries provide a consolidated approach in line with smart city vision and objectives, effective resource allocation, reallocation, and distribution to establish initiatives and proposing innovative approaches to optimize a city's human and financial resources, as well as regulating planning capacity and knowledge.

Under the Smart City Mission, the Command and Control Centre (constituted as an SPV) is a centrally coordinated system established to collect information across the smart city area. In Bhubaneswar, there is also a BUKC under MoHUD to coordinate data collection for key department and stakeholders. It is clear from the survey results that these institutional mechanisms are critical for effective and efficient city management and development.

On the governance level, administrative policies and availability of required resources were highlighted as an issue which affected project implementation time and service delivery. Political interference, obtaining adequate resources and delays in approval times were also singled out as key challenges. A major problem referred to is inadequate inter-departmental and agency coordination due to bureaucratic silo effects. In this case, the BUKC type of mechanism that applies the existing knowledge base and ensures close coordination with engineers and urban planners in different departments, including resource allocation, can make a difference.

These central units at national level are often temporary measurement facilities and are eventually phased out as part of the exit strategy of the Smart City Mission. But reestablishing the centrally coordinated system under city authorities, such as the Bhubaneswar Development Authority, the SPV, becomes an effective exit strategy. Cities need to employ their own local mechanisms with clear descriptions of roles and responsibilities assigned to stakeholders, key ministries, and agencies.

These are several examples that can also be applied to urban development. In smart and resilient cities, it is crucial to prioritize addressing fundamental gaps and achieving a balance in providing diverse services through technology and an enabling environment, rather than adopting technology in isolation under the label of a smart city.

6.5 Early Integration of Soft Assets

This research reinforces the value proposition that using soft assets to create an enabling environment in the early planning stages is critical for developing effective smart, resilient cities. So far in this chapter, the following points were discussed.

(i) Identification of soft assets influential nodes can be valuable information to facilitate the strategic incorporation of soft assets and understanding the value creation process.
(ii) Soft assets support the development of enabling factors for quality service delivery, and the level of soft asset consideration and utilization can contribute to a city's EOL and MPI results.
(iii) There are inherent challenges in using soft assets and, in particular, political influence in the public sector presents additional difficulties. Therefore, to effectively create an enabling environment for soft assets, it is important to understand how to use soft assets effectively to fill gaps and achieve desired outcomes. This is the starting point for measuring soft asset outcomes.
(iv) The value creation mapping process is dynamic and organic. It is important to examine the process both at project and city level, as the sum of individual results alone will not capture the city environment as a whole. A separate value creation mechanism is needed and effectively integrated into the planning and development process.
(v) Soft assets are also knowledge assets. It is an important aspect in terms of citywide learning and productivity. This also corresponds to the public sector value creation framework examined in Chap. 3. Also, soft assets as knowledge can contribute to optimal outputs.
(vi) With the many challenges in developing countries, adding a central coordination mechanism can optimize the benefits derived from 'smartness'. However, implementing cross-departmental mechanisms remains inherently difficult in the public sector. In the smart, resilient city context, central coordinating mechanisms require technology, data-driven decision making, public involvement, and innovation as a major focus. This leads to an effective way to create an enabling environment which provides a consolidated approach, effective resource allocation, distribution and reallocating initiatives, and adding innovative approaches to optimize a city's human and financial resources, regulating planning capacity, learning and knowledge sharing.

One of the key recommendations of this research is the importance of early integration of soft assets in the project planning and implementation stages and as well as a comprehensive application citywide. Soft assets must be introduced in advance of the planning and implementation of hard infrastructure to fill gaps and provide necessary solutions. Incorporating soft assets at the early development and planning stages increases: (a) the optimization of technology use, and (b) timely use of resiliency solutions for external shocks. Some results in the city surveys reinforced these points. Hence to optimize efforts to include soft assets, it is important to understand the value creation process and incorporate them early into project planning and policy making.

As for the optimization of technology use, this research found that soft assets can enhance enabling factors and improve results for a city in EOL and MPI evaluations. Also some future impacts and outcomes can be anticipated from current value creation mapping. On the latter point, incorporating soft assets in the prevention and evacuation stages of planning increases resiliency toward external shocks. The research outlined in Chap. 5 about the level of vulnerability and capacity of Bhubaneswar (Kawyitri and Shekhar 2021) strongly supports the importance of early incorporation. Social and human factors can minimize the effects of disasters like cyclones compared with other physical, smart-city and financial dimensions. It takes time for a city to put soft asset elements, such as communication techniques in place. It is demonstrated that a strong community response and evasive action from citizens create more self-dependent and adaptive defenses against disaster.

These two points exemplify the intricacies involved in the process of creating value from soft assets. The results of the level of vulnerability and capacity assessment in Bhubaneswar clearly demonstrates that the value creation process occurs at multidimensional layers of the city administration. As mentioned earlier, the process evolves organically and is influenced by unpredictable factors as value is derived from the specific context. The value created evolves or disappears over time depending on the prevailing circumstances. This further emphasizes the importance of early consideration and utilization, as well as assessing values at the horizontal or comprehensive city level.

6.6 Comprehensive City Level Appraisal

Chapter 1 makes the case that the theories of living structure and wholeness can be a useful marker for the importance of early soft asset integration and city level evaluation.

The idea of 'Living structure' was developed using the notion of 'beauty'. Despite the different targets the apprehension of 'beauty' and soft assets, both largely depend on subjective judgment, but also objective dimensions. With this notion, it is possible to explore the complex process of capturing soft asset value creation and appraisal using the notion of 'wholeness'. Using a computation approach a 'notion of beauty' is used to analyze one 'beauty' image by classifying many small items into the inherent scaling hierarchy. This research employed this analysis to classify many small items or nodes to isolate the influential node and pinpoint the entry point of the value creation mechanism. Soft assets value creation involves multiple stakeholders and different contexts which is similar to the research associated with the nature of 'beauty'. The key idea of wholeness creating beauty also involves a dynamic and organic process and, like soft assets, difficult to describe and assess.

The important findings from this research are: (i) the key elements of value creation can be depicted in the network analysis through isolating influential nodes supported by small nodes. However, the actual process is intricate and influenced by unpredictable risks and specific values evolving from an organic process related to the specific context; (ii) the process itself needs to be incorporated by cities into the smart, resilient city development process to create an enabling environment at early stages; and (iii) the process can be utilized to develop an approach to measure the effective integration of soft assets.

6.7 Co-value Creation

Based on the initial discussion in Chap. 3 and the foundational arguments outlined in this chapter, this sub-section describes a soft assets co-value creation model for use in smart resilient city development.

In Chap. 3, the public sector value creation model of Kaplan (1999) was examined. Also discussed was the importance of establishing a two-way interaction between the city service provider and the citizens or consumers receiving the service delivery. Based on the literature review and the survey results, a credible model can be developed to provide a framework to better understand the relationship between soft assets and the value creation progress.

Figure 6.3 shows a reciprocal model of co-value creation. It places weight on the value of 'quality communications' and relationships between city administrators and community beneficiaries that will ultimately encourage hard infrastructure investment and creating an environment for learning and growth in each of the two tiers. Balanced interaction on both sides of the equation– providers and recipients—creates the desired outcomes and the value necessary to lay the foundation for sustainable smart urban development and resilience. The important point here is that both ends of the spectrum have to develop the capacity to create value within each group.

On the left side of the diagram, there are opportunities for value creation where city authorities deliver services to citizens. The quality of service delivery can be enhanced by incorporating soft assets. Also reliable capture of accurate internal knowledge, learning, productivity and growth strengthens and improves the process. The effectiveness of the way the city authorities use soft assets in the initial planning and implementation stages contribute to the overall improvement of the three higher level objectives—'cost of providing services', 'value/benefit services' and 'support of legitimizing authorities' which support a city's service delivery providing finance and other means (citizens, etc.). In essence, it emphasizes the value of internal learning and growth within city administrations.

The right side of the diagram represents the community at large where citizens participate—'consuming services', 'providing performance 'feedback system

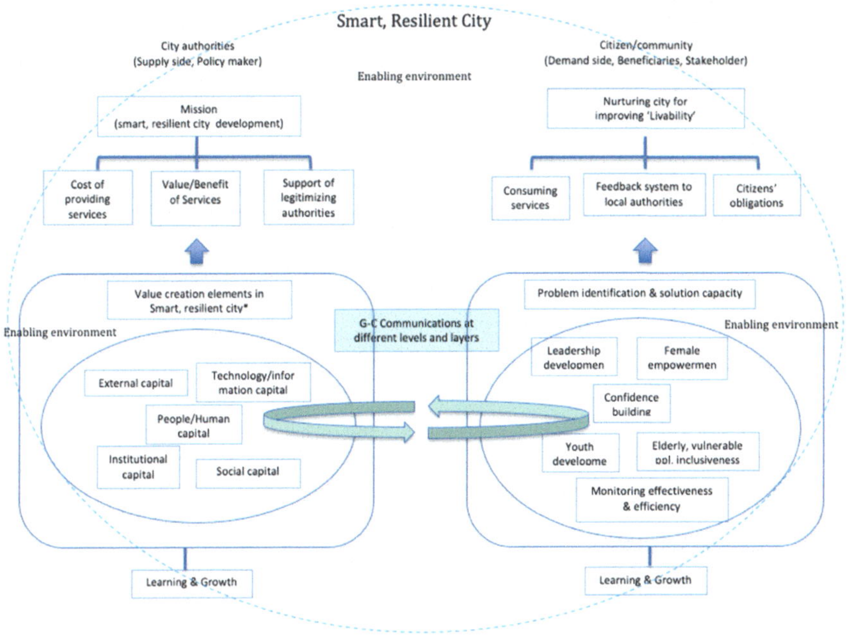

Fig. 6.3 Reciprocal model of co-value creation (*Source* Wataya 2023)

to authorities', and 'citizen's obligations, including paying tax. Some of the findings in this research illustrates the two sides. Outcomes will depend on different contexts, including the capability and knowledge of the local community. In this research, the listed items are used as examples. They basically contribute to building capacity and creating fundamental assets of value to the urban community, which is at the core of value creation, promoting internal learning and growth within communities.

This figure also explains the unique aspects of knowledge as a soft asset. Soft assets are related to knowledge and strengthened by external relations and synergetic actions which comprise various human experiences motivated by different desires, objectives or incentives. At both the administrative and community levels, the body of knowledge and experience belongs to each individual, not corporate assets owned by the city authority, entity or community. Furthermore, the collective body of knowledge and information needs to be regularly updated in the smart and resilient city development process. In this regard, the capacity of the personnel in smart city authorities and the knowledge of citizens in the community are essential to create broader social capital for long-term development and sustainability. This 'two-tiered nature' of people/human involvement creates another level of complexity in the value creation process.

Quality outcomes depend on better interactions a two-way relationship and communication at all levels within the city administration and across the urban community under the city's jurisdiction. As the survey results demonstrate, using the consultation process and engendering effective communication is one of the most successful approaches to achieve positive outcomes. For the city to fully benefit from the feedback and produce better outcomes, it is highly important to undertake quality, in-depth consultations and follow-up on a continuous basis. Ultimately, these sustained actions develop an enabling environment and a foundation in the city that increase the effectiveness of urban development projects. This research and the underlying survey data indicate that the reciprocal model of co-value creation also operates effectively as the value creation process.

References

Kawyitri N, Shekhar A (2021) Assessing vulnerability and capacity of Bhubaneswar as a progressive smart-city: An empirical case study of Fani cyclone impact on the city. Int J Disaster Risk Reduction 56(1):101986, https://www.sciencedirect.com/science/article/pii/S2212420920314886

Pagano and Neubert (2015) Building resilient SCs: mapping and measuring key value drivers through a multi-perspective and multi-stakeholder value creation framework based on intangible assets, Conference paper, Part of the IIP advances in information and communication technology book series (IFIPAICT, volume 463)

Tan YS, Taeihagh A (2020) Smart city governance in developing countries: a systematic literature review. Sustainability 12:899. https://doi.org/10.3390/su12030899, 1–30 https://papers.ssrn.com/sol3/papers.cfm?abstract_id=3526607

Wataya E (2023) Integrating soft assets in smart and resilient city development in India. Unpublished PhD thesis, Keio University, Japan

Chapter 7
Conclusion

The previous chapter discussed key findings from the survey, namely the identification and verification of influential nodes, including factors contributing to the livability index, challenges in establishing smart, resilient cities, and the importance of early integration of soft assets to create an enabling environment for smart, resilient cities. The value creation process is a key mechanism using technology which enhances 'smartness' to build resilience and sustainability. Combining the literature review with the analysis of the survey results of this research, a key additional finding is that the value creation process is complex, evolutional and occurs across several city dimensions and layers.

Also reviewed was the "Blackbox" concept and value creation mapping. As the process evolves over time, various interconnected soft assets are employed, representing the anticipated and actual outcomes of a project with either value created or a potential loss of opportunity. The simplified "Blackbox" analogy with the value creation mapping tool not only computes the project's specific objectives and tangible outputs, but also underscores the benefits which accrue to the city as a whole. The evidence is clear that soft assets provide an important foundation when cities design and implement infrastructure and technology to create a smart resilient city environment.

In city development, the process of reciprocal co-value creation is an internal process depending on the quality of external relations between the city authorities and the community, as well as other external and unpredictable factors. At the same time, a value creation model indicates that the availability of time is a key factor leading to the early integration of soft assets. Another unique feature is that city authorities and the communities need an effective enabling environment and adequate capacity throughout the value creation process. A two-tier co-value creation model is proposed, relying on the value of quality communications and creating an environment for learning and growth using human capital, effective leadership, and fostering innovative systems and sustainable development. Within the domain of human capital, there are two-different elements: the owner of a body of knowledge

E. Wataya and R. Shaw, *Soft Assets in Smart and Resilient City Development*, Disaster Risk Reduction, https://doi.org/10.1007/978-981-97-3373-6_7

and soft assets embedded in the systems and processes of the city administration. Moreover, the value creation process itself becomes a soft asset and occurs at different levels: individual, institutional and citywide.

The research demonstrates that the consideration and use of soft assets is vital for the development of smart and resilient cities. Taking into account the survey results and the discussion and analysis set out in the previous chapters, there are five key findings discussed below.

7.1 Summary of Key Findings

1. Enhancing soft asset incorporation: Establishing a smart, resilient city is one of the major solutions to address the challenges caused by rapid urbanization in developing countries. A robust enabling environment establishes factors essential for realizing the vision and goals of a smart and resilient city. Vital actions and inputs, such as raising awareness of soft asset benefits and uniqueness, are necessary not only for city administrators, but also for other stakeholders, including citizens. Specifically, a smart city should enhance service quality using technology to promote more effective and efficient use of resources. This research explores the equally important role of soft asset capital. Without a technologically favorable environment and a range of soft assets, a city cannot unlock the full benefits of using technology. This hinders the development of sustainable smart and resilient cities. A major finding of this research is that project implementation in smart cities requires the balanced use of both hard infrastructure paired with soft assets to achieve long-term sustainability.

The research identified multiple challenges faced by cities in developing countries to implement smart solutions in the context of rapid urbanization. The survey results pointed to the lack of enabling factors in the cities to optimize 'smartness'. This finding corresponds to another study detailing challenges in developing countries for smart city development (Tan and Taeihagh 2020). The survey results also highlighted the challenges about governance, administrative policies, the quality of technology and availability of human resources, often filled by the private sector. The principal challenges are a deficient enabling environment to implement the necessary changes. This makes it difficult for cities to fully benefit from smart policies and solutions. Also, the findings and analysis in Chap. 5 showed the importance for cities to deliver basic services by infrastructure. However, soft assets are rarely considered by city administrators and cities have substantial challenges using soft assets to supplement hard infrastructure approaches.

With the rapidly growing global trend towards building smarter and resilient cities, the application and incorporation of soft assets are critical for delivering infrastructure to provide basic services and to create long-term value and urban sustainability. Although there is no one size fits all path to smart, resilient city development, the research has identified three key approaches:

(i) a strategic approach: Although there is an early emphasis on strategic and systematic consideration of soft assets, they are often regarded by city administrators as implicitly included. The soft asset capitals examined in this study, for instance, are integrated and included at the strategy and project planning stages to some extent. However, in many cases, there appears to be a lack of a well-established system for the efficient utilization and management of soft assets. For example, the survey responses mentioned human resources, specific capacity development, including on-the-job training and the involvement of external technical experts as crucial for project implementation. There needs to be a more cohesive strategy by the city to embed enabling factors, essential to integrate soft assets at an early stage. Chap. 3 explained that in each particular city context, soft asset capital components drive specific value. The review of the literature and previous research confirmed that soft assets create a solid foundation to operationalize basic infrastructure and organizational structures that benefit citizens and the community as a whole. However the unique and intangible characteristics of soft assets can act as a disincentive for their utilization. As argued in Chap. 5, value creation or the benefits of soft assets accrue from consideration and incorporation at an early stage of project development.

(ii) a balanced approach to integrating soft and hard assets is essential. Achieving an optimal balance is critical for achieving the unique development objectives of each city. Achieving a realistic balance is difficult due to the inherently complex factors involved, differing priorities and circumstances, and the need for constant adjustments on the ground. A city faces substantial challenges in achieving this balance. The survey results reveal other priorities which favor the visibility provided by physical infrastructure, potential political gains for improved performance rankings and insufficient research and data about the benefits of soft assets. This results in challenges related to accurate estimation and resource allocation, as well as lack of departmental and inter-agency coordination. These issues are not exclusive to developing countries; they also apply to the public sector in many countries and represent persistent challenges that impede the pursuit of a balanced approach.

The India Smart City Mission, while heavily focused on infrastructure construction, also focuses on the need to improve living quality, which affects the responses about the challenges of soft asset incorporation. The type of services and products developed under the Smart City Mission has the potential to improve data quality, collection and analysis. A phased planning and intervention strategy, led by city authorities with effective leadership, involving people-centric decision-making, can contribute to creating a more favorable environment. Leveraging global best practices and actively incorporating national-level initiatives and directives are considered crucial initial steps to influence both planning and implementation processes.

(iii) institutionalizing soft assets management mechanisms to achieve long-term sustainability: This issue extends beyond smart and resilient city development. As outlined in Chaps. 5 and 6, recognizing the diverse roles and responsibilities of city-level institutions is crucial for efficient resource utilization and effective city operations. From the survey results, institutional mechanisms, such as establishing a central coordinating function, a special purpose vehicle or a centralized coordinating entity, can not only play a key role in planning comprehensive urban development, providing urban services and bridging communication gaps between different agencies, but also in building a strong foundation for retaining knowledge and establishing best practices to strengthen institutional capacity. Using a central coordinating mechanism assists to comprehensively integrate soft assets and strengthening the enabling environment for smart, resilient city development. Bhubaneswar provides an excellent example of this outcome. Functions, such as providing useful solutions, data-driven decision-making, research, and knowledge dissemination, are separate, but closely connected in the survey and present opportunities to address gaps on the ground and facilitate the integration of a soft asset value creation process.

2. Assessment framework: Utilizing soft assets are vital to creating sustainable urban solutions. Each category of the five soft asset capital components cover a wide range of items. They are not absolute and need modification to better meet the local context for creating smart, resilient city solutions. The same approach can be applied to the definitions of the nine key elements specifically selected for this research. These elements are tailored for this research and applied to the smart and resilient city context referring to existing indicators and major pillars. It also emphasizes that the elements are not only related to technology and adaptability, but also to key features unique to smart and resilient cities. Customized definitions can further help to identify soft asset integration.

Chapter 3 set out an in-depth literature review of research on intangible assets. To make smart cities resilient and sustainable, it is argued that "value creation" through soft assets is essential for cohesive smart city development. Soft assets create an enabling environment in cities for operationalizing basic infrastructure and the organizational structure required to produce social benefits for citizens and consumers. Here, spillover, synergistic, and scaling-up effects create value through a dynamic and organic process.

Defining and identifying soft assets was explained in a broad context. This is because building a solid foundation to accomplish city authorities' goals and responsibilities and implementation of public works requires dealing with a range of issues and different types of soft assets embedded in smart and resilient city key elements. These include enabling soft assets like human considerations, a wider range of systems from political, analytical, planning, and accountability, to technology and information. External relations is also a critical component of soft assets. Social capital is particularly difficult to capture due to more limited information available, but remains important in this research as it constitutes enabling capital required to operate communities effectively. It also assists communities to learn and grow. Although the five capital components cover a wide range of items, they are supported by a substantial amount of previous research and studies. The soft asset capital components selected can also be compared to the way private firms manage intangible assets. Developing specific and clear definitions of these soft assets can further help identify how soft assets can most effectively be adapted to the development context.

Because the specific value of soft assets depends on the local city context, the research also identified nine soft asset core value creation elements representing the components of smart and resilient city development. These nine value elements are unique to this research, but not all encompassing. Each one is different and broadly cover: (i) the key features of smart, resilient cities; (ii) different smart city approaches; (iii) the three higher level objectives of the value creation model assuming that the soft assets drive value creation. There can be more than nine elements, but the nine selected are sufficient to cover the key clusters described in this research. Based on the above nine soft asset value elements and the five soft asset capital components, a qualitative framework has been developed. The purpose of this tool is not to evaluate soft asset performance when smart city initiatives or projects are being discussed, but to assess the level of utilization of soft assets in line with the overall vision and objectives of the city.

The uniqueness of the nine value elements can be used as one example to demonstrate a combination of bottom-up and top-down smart and resilient city approaches described in Chap. 1. Other smart city classifications include community engagement, communications, and social capital aspects, but they comprise part of a larger set of pillars representing broader city characteristics. For example, Giffinger et al. (2007) presented a hierarchical structural analysis consisting of 6 characteristics explained in Chap. 3, with 31 factors and 74 indicators, which apply to and rank medium-size smart cities in Europe. However, elements relating to community engagement, communications, and social capital aspects are combined with other main pillars; for example, community, and aspects related to efficient city government administration are classified under 'smart governance' and 'smart people'. Elements involving participation in decision making and political strategies are covered under 'smart governance'.

'Resilience' is a concept that can cut across other value elements and stand alone. This challenge lies in effectively incorporating resilience aspects distinct from other elements. The tool developed and employed in this research utilized a combination of different categories. Streamlining these variations is necessary for the practical application of the tool. This also applies to the concept of 'social capital', which is a socio-cultural asset within the five soft asset capitals. Notwithstanding this analysis, the proposed tool elaborated in this research can be used in a unique way to review city administration efficiency and the community response, as well as the key pillars required to establish a smart, resilient city.

It should be reiterated that the qualitative tool for measuring soft asset consideration has some limitations due to its purpose and the classification approaches mentioned so far. Inevitably, the qualitative tool subjectively analyses and classifies data based on individual knowledge and experiences. To ensure the tool can become a more objective means of evaluation, it is necessary to reduce the subjective judgements associated with individual knowledge and experiences. One way to achieve this result is to further subdivide and adapt both the classification of the five soft asset capital components and the nine value creation elements to local city contexts.

3. Analysis of soft assets incorporation: A comparative analysis of smart cities shows consistency between proposal and implementation stages and also correlation with livability indexes. The qualitative tool used to identify soft assets integration is one way to further develop measurement and assessment of soft assets value. Incorporating soft assets in this context demonstrates their alignment with the vision and objectives of smart and resilient city development, emphasizing that effective integration is essential for project success.

Chapter 5 described a network analysis and value creation mapping of soft assets in each city showing how small nodes (initiatives, activities using soft assets) link to a large node (key core value creation elements). Also this chapter examined how the different approaches of the two cities selected for case studies determine how soft assets were incorporated into their planning and operations. Based on the literature review and this research, influential nodes in the network have many smaller nodes. While the influential node may not demonstrate actual performance, many of them align with the survey results, suggesting their substantial contribution to overall performance. Small nodes are more important than larger ones, working synergistically, with spill-over, and scaling-up effects. However, tracing their impact is not feasible due to the nature of soft assets, and, as discussed in Chap. 5, is not particularly meaningful.

The analyses were validated by field investigation and in-depth interviews conducted with city administrators, SPV staff, private sector firms, non-profit organizations, and academia in each city.

Value creation mapping exercises based on the survey responses also indicated similar results for the network analysis. The value creation mapping is, however, still rudimentary and difficult to follow up, but it is nevertheless a useful tool for policy makers, decision makers, and stakeholders to understand how the value of soft assets can be appraised. The mapping exercise highlights the importance of time as a factor which affects inputs to output, outcomes to impact.

The analysis shows that in case of Bhubaneswar, the applicable influential nodes are identified as 'quality communications', 'community engagement', 'smarter services via technology', and 'equal access to services'. Among these, value creation is most evident in 'quality communications' and 'community engagement'. While in Nagpur, the focus is on: 'equal access to services' and 'smarter services via technology', 'health, well-being', and 'resource management'. Among these value creation elements, the most prominent is 'smarter services via technology', and 'equal access to services'. Desk top research and the results from the network analysis demonstrated that the degree of soft asset integration were different in Bhubaneswar and Nagpur. Each city's approaches were aligned with the city's particular strategy and vision.

It was also found that higher EOL and MPI scores broadly corresponded with the desk top network analysis and the survey results regarding the degree of soft asset incorporation. This indicates that the degree to which the city considered soft assets affected the level of EOL and EOL scores. MPI sector verticals with higher scores also corresponded to the influential node. The robust planning and institutional capacity of Bhubaneswar is an enabling factor which has positively affected its smart, resilient city planning and implementation. Bhubaneswar's outcomes were verified by its high rating in the Citizen's Perception Index. Nagpur employed more robust technology approaches in its smart city proposals and implementation stages. The obvious development impact and values were created by effectively using technology to deliver effective services. This was also validated by Nagpur's performance score in 'service' which is higher than the national average.

There is a strong correlation between the results from the proposal stage with those of the field survey. Increasing incorporation of soft assets in the planning stage in a specific area corresponds to a similar incorporation during implementation with clear results. In situations where an effective plan is put into action, it may deviate from the initial plan or, once implemented, it could result in additional improvements. This in turn, reflects the city's broader development focus. While this research does not provide a definitive answer, it shows that incorporating soft assets with hard infrastructure is better aligned with the vision and objectives of a smart and resilient city. This analysis and the tools developed help the preliminary assessment of the classification of the pillars necessary to incorporate soft assets.

The 'consultation' and 'Learn and Share' (L&S) approaches can provide solutions for challenges in developing country cities where soft asset capital provides a unique contribution to city innovation and value. L&S will also become useful to document and learn from failed cases classified as 'value not created' to convert the failures to positive results over time. To understand these unique city values, it is important to review how the 'Learn and Share' process is used, including learning from experiences, successes, failures, and the internalization process within each city administration.

4. Resilience: Resilience is also supported by robust infrastructure, data and technology, community-led resilience initiatives, adaptation to environmental issues, sound leadership and institutional capacity. As resilience capabilities progress and are integrated into the smart city development process, it adapts, adjusts and contributes to the dynamism of the smart city development process. Given the heavy reliance on soft assets, early and strategic incorporation is crucial for building solid systems and social infrastructure.

India has implemented one of the most extensive national smart city programs to deal with the challenges of rapid urbanization by modernizing its infrastructure and improving citizens' quality of life. A convergence scheme also fills the gaps in services provided by using the Smart City Mission resources. Basic needs provided by core infrastructure were incorporated into the Smart City Mission which improved the livability and sustainability of the cities. Soft assets are proactively incorporated in this process and, as the survey results showed, much can be learned from this experience.

However, despite the high risk of damage and destruction from disasters, mechanisms for disaster risk management and climate change adaptation were not originally included in the Smart Mission proposal. Action needs to be taken to address long-term impact in addition to focusing on immediate prevention and evaluation measures. Consequently, the new framework provides for climate assessment and mitigation to complement the Smart City Mission: The Climate Smart Cities Assessment Framework 2.0, which covers 5 key pillars to remediate climate change issues and environmental degradation (outlined in Chap. 2), is an effective tool to complement the objectives of the Smart City Mission. Creating feedback loops to improve the indicators and assessment methodology, is critical to assist in the integration of comprehensive climate responsive development.

Resilience is supported by robust infrastructure, competent leadership ('people/human'), sound institutional capacity ('institutional'), reliable data and technology ('technology, information'), quality communications ('external'), community resilience ('social') and effective adaptation to the environment. As the resilience initiatives are integrating into smart and resilient city development, it adjusts and strengthens through an organic process, changing, transforming and absorbing better inputs from innovation, experiences, information, knowledge and learning within the overall smart and resilient city development process. Early incorporation of soft assets play a significant role in strengthening social infrastructure and systems to build resilience.

The devastating 2020/21 floods in Nagpur and the 2019/20 cyclones in Bhubaneswar delivered important lessons. Nagpur's approach to disaster resilience focused on using smarter technology in sharing flood information, while Bhubaneswar focused on inclusive approaches, working with people with different roles and responsibilities, including community-based preparedness and local resilience initiatives. In Bhubaneswar, vulnerability and capacity assessment research (Kawyitri and Shekhar 2021) indicated that social and human dimensions were more significant than infrastructure, smart technology, finance, and other hard aspects. The analysis of the contribution of the India Smart Mission program to this capacity development validated the network analysis in this research. Using the influential nodes analysis, focus areas identified in each city are aligned with each city's smart city development approaches. However, this is only a snapshot. To address current issues and gaps, strategic consideration of soft assets would be an important entry point to improve disaster risk management.

From the literature review and the survey responses, the gaps between immediate and long-term impact can be addressed using a better balance of soft assets and hard infrastructure, including communications, awareness, behavioral changes, regulation, cross-departmental coordination mechanisms, as well as reliable social, health, financial, and other services that empower individuals to be self-reliant and recover quickly in the aftermath of the disaster. In this context, soft assets can play a strong role in improving urban resilience.

The main conclusion is that early incorporation of soft assets in the planning process is important to build resilience. Referring back to the "Blackbox" the collection of individual values are not equated to overall city values. The 'Blackbox and 'value creation cases' covered in Chap. 5 was simplified to present the flow from inputs to outcomes over time. The various activities examined in Chap. 5, demonstrated that soft assets create value invisibly using the "Blackbox" analogy in Chaps. 1 and 6. Soft asset contributions to outcomes in early stages are generally lower, but gradually increased over time. Over time, it becomes more challenging to clarify the link between inputs and outputs to outcomes, and the creation of value. Similarly, understanding the development of resilience over time becomes more complex. Nonetheless, value creation takes place when the generated outcomes deliver benefits or advantages. As demonstrated regularly in this research, the value creation process is dynamic and evolves organically, and is also affected by unpredicted factors and risks. It is important to incorporate this dynamic process into city planning and implementation, but better outcomes occur when this is done as early as possible. In this process it is also important to take into consideration failed initiatives or actions considered 'not value created' as discussed in Chap. 6.

5. Tools and models: A qualitative framework and reciprocal co-value creation model could be one useful entry point to measure how soft assets can be more effectively incorporated into the planning and implementation stages of smart, resilient cities.

The survey results set out in Chap. 5 suggest that cities trying to incorporate soft assets integration can easily be impeded by political bias favoring hard infrastructure which produces faster, more tangible and measurable results. It is important, therefore, to develop a mechanism to review, identify gaps, monitor, and evaluate performance associated with smart resilient city vision and objectives.

The city's capacity to create an effective enabling environment is fundamental in developing a smart, resilient city. These observations can also be applied beyond smart and resilient city concepts to every aspect of city administration and management, but the lack of enabling factors diminishes the city's ability to apply technology which delays development and deprives the benefits of 'smarter smartness' in the case of adapting resilience to external shocks. This is a significant challenge for most cities in developing countries.

Developing a tool or analytical framework to understand the complexity, unpredictable nature, and wholeness of soft assets is an important building block towards creating smart, resilient cities. The tool can be highly useful to identify and understand key aspects to be examined and measured before the next stage of planning, implementing, and management of soft, intangible assets. IoTs, or technology integrated infrastructure incorporating soft assets, can effectively be used to assess impact and results. Improving soft asset management and developing effective measurement tools can strengthen the capacity of a city to deliver quality services to its citizens. At the same time, citizens also need to improve their awareness, knowledge and capacity to consume services, provide useful feedback to the city and fulfill their obligations, including paying taxes for service provided (outlined in Chap. 3).

As described in Chap. 6, the process leads to the co-value creation model that functions and interacts at city and community level respectively. This research used the survey results to identify six specific components underlying the creation of a values-centric environment for the community: (1) leadership development, (2) women empowerment, (3) confidence building, (4) youth empowerment, (5) inclusivity, including accounting for elderly and vulnerable people, and (6) enhancing effective monitoring. This also implies an effective learning and growth process within the community, with similar actions by the city administration. A co-value creation model can be utilized to proactively integrate the components at different levels and achieve scalability. The components can be reviewed and changed to reflect the local city context. This creates a dynamic, continuous learning process between the community and across and within the various levels of the city administration. The learning process itself will also constitute an effective soft asset utilized by both the city institutions and the community.

7.2 For Future Study

Based on the findings of this research, the following areas are briefly suggested as future directions for study and research on this topic of integrating soft assets in smart, resilient city development.

1. Refinement of the Qualitative framework to identify and adaptation of soft asset incorporation

Creating mutually exclusive indicators in the tools proves challenging, given that the design primarily caters to qualitative assessments. During the desk top research and survey result assessment process, inevitable overlaps occurred due to the tool's inherent nature. However, potential enhancements could involve further subdivision into detailed sub-categories for the nine soft asset value elements and five soft asset capitals. This could include further categorization of the various types of pillars or isolating specific pillars to develop more in-depth tools for practical applications.

2. Further verification of value creation mapping results

The research has limited information to create the value creation mapping. Also detailed information about time factors were not fully incorporated. The development of more detailed and precise value creation mapping can not only help city policy makers and administrators but also various stakeholders better understand the importance and priority of utilizing soft assets to derive benefits from the elements of city 'smartness'.

3. Integrating the value creation process into the city development (practical approach)

The research discussed the importance of integrating the value creation process in various city development stages. There are already different approaches such as characterizing soft assets as knowledge assets, institutionalization of learning and growth or learning and sharing. For instance, given the need to strengthen disaster risk management in the Smart City Mission Program, the approaches to identifying soft assets incorporation needs can be further customized for disaster risk reduction and management as well as developing associated indicators by separating 'resilience' elements from the qualitative framework referred to earlier. Further research and analysis of the details of the disaster risk reduction and climate change adaptation programs of the Smart City Mission Program can provide more insight about how soft assets can strengthen city disaster risk management and resilience initiatives.

4. Measuring soft asset value at city level

Individual value creation requires analysis and evaluation to identify, implement and support the utilization of soft assets. However, since the soft assets value creation process is an evolving phenomenon, further study at the city level value creation approach could pave the way to integrate soft assets at various levels of city activity. This includes further development and refinement of the qualitative tool and co-value creation appraisal methods at city level. Using the MPI index for evaluating municipal performance on enabling factors, for example, could be a useful tool to study further how to mainstream soft asset capital components into smart, resilient city development strategies.

References

Giffinger R, Fertner C, Kramar H, Kalasek R, Pichler-Milanovic N, Meijers E (2007) Smart cities, Ranking of European medium-sized cities. Final report, October 2007, Delft University of Technology, https://www.researchgate.net/profile/Rudolf-Giffinger

Kawyitri N, Shekhar A (2021) Assessing vulnerability and capacity of Bhubaneswar as a progressive smart-city: An empirical case study of Fani cyclone impact on the city. Int J Disaster Risk Reduction 56(1):101986, https://www.sciencedirect.com/science/article/pii/S2212420920314886

Tan YS, Taeihagh A (2020) Smart city governance in developing countries: a systematic literature review. Sustainability 12:899. https://doi.org/10.3390/su12030899, 1–30 https://papers.ssrn.com/sol3/papers.cfm?abstract_id=3526607

Definitions

Betweenness Centrality of a node

- The amount of control that one node exerts over the interactions of other nodes in the network

Blackbox

- In this research, Blackbox is a hypothetical model suggests non-obvious causal relations and linkages between inputs and outputs.

Co-value creation model

- Co-value creation model is a proposed model relying on the value of quality communications and creating an environment for learning and growth using human capital, effective leadership, and fostering innovative systems and sustainable development.

Integrating (or incorporation of) soft assets

- The decision-making process by city administration and officials to use soft assets in project planning to implementation stage. In this research, "integrating soft assets" covers the following: (i) developing enabling factors and a favorable environment to utilize soft assets; and (ii) using soft assets to operationalize projects; (iii) creating value using components of soft assets. A main assumption of this research is that: (a) integrating soft assets is more effective earlier in the planning and implementation process; and (b) this value creation process should be internalized in the development process to build smart, resilient cities.

Influential (large) nodes

- "Influential (large) nodes" are soft assets components that create value and more commonly utilized than other elements. Large nodes are composed from smaller nodes (represented by activities, initiatives) that are associated with elements of soft assets. These nodes are identified by network analysis.

E. Wataya and R. Shaw, *Soft Assets in Smart and Resilient City Development*, Disaster Risk Reduction, https://doi.org/10.1007/978-981-97-3373-6

Qualitative framework for identifying soft asset consideration

- To identify and analyze the process of value creation within city development activities, a soft asset identification tool with 45 items was developed based on nine (9) elements and five (5) soft asset capital components. Further details are explained in Chap. 3 and the results of the use of the tool is set out in Chap. 5.

Small node

- In this research, small node represents a separate activity associated with nine vale elements.

Soft (intangible) assets

- "Soft assets" can generally defined as non-physical, but identifiable assets. Some assets can be identified directly and valued, but some assets are intricately linked to other components, including hard infrastructure and more difficult to distinguish and quantify. Often there is no clear definition and classification of soft assets and their value depends on the specific context. Thus the concept and definition of soft (or intangible) assets embrace a broad range of areas such as human resources, policy application, institutionalized processes, stakeholder networks, data, IT applications, communication, outreach public relations, community level organization, interactions with representative groups, etc. This research makes the claim that soft or intangible assets are equally important within city planning and implementation as hard or physical infrastructure to deliver quality services to citizens.
- In this research, soft (or intangible) assets refer to the following: (a) People/Human capital, (b) Institutional capital; (iii) External capital; (iv) Technology, information capital; and (v) social capital. Further details are described in Chap. 3.

Soft (intangible) asset capital

- There are several variations of the term "intangibles", i.e. intangibles, intangible assets, intangible capital, intellectual assets, or soft assets as opposed to hard assets. In this paper, the term intangibles and soft assets are used interchangeably.

Value elements of Soft Assets in smart and resilient cities

- Due to the characteristics of soft assets (synergetic, spillover, scalability, context specific value, etc..), this research identified 9 "value elements" in smart and resilient city. These elements are core pillars of smart, resilient city development specific to this research purpose. They are: (i) resource management: (ii) accessibility (mobility); (iii) health well-being; (iv) equal access to services; (v) smarter services; (vi) community engagement; (vii) quality communication; (viii) security; and (ix) resilience. Further details are explained in Chap. 3.

Wholeness

- Christopher Alexander's work refers to creating built environments that are coherent, harmonious, and supportive of human well-being. In this research, this idea was referred as creating environments that feel complete where the elements work together in a way that enhances the overall experience and functionality of the space. It encompasses the integration of various elements to create space and value.

Bibliography

Annual Percentage of Population at Mid-Year Residing in Urban Areas by region, subregion and country, 1950-2050, https://population.un.org/wup/Download/, United Nations Department of Economic and Social Affaires Population Division)

Bhubaneswar Development Authority, https://www.bda.gov.in

Bhubaneswar India Local Governance Network, https://www.preventionweb.net/files/globalplatform/entry_presentation~13h00.pdf

Bhubaneswar Municipal Corporation, https://www.bmc.gov.in/departments/disaster-management

Bhubaneswar Municipal Corporation, https://www.bmc.gov.in/about/profile

Britannica, https://www.britannica.com/place/Bhubaneshwar

Bhubaneswar Smart City Limited, https://www.smartcitybhubaneswar.gov.in/know-your-bhubaneswar

Consultative Group for International Agricultural Research (CGIAR), Resilient Cities, https://www.cgiar.org/initiative/16-resilient-cities-through-sustainable-urban-and-peri-urban-agrifood-systems/

District Disaster Management Authority, Nagpur, Nagpur government, District Disaster Management Plan, Nagpur District, 2017–2018, https://static.s3waas.gov.in/s3d1f491a404d6854880943e5c3cd9ca25/uploads/2018/03/2018031651.pdf

Disaster Management Unit, Relief and rehabilitation Department, Government of Maharashtra, "Maharashtra State Disaster Management Plan' Draft copy (2015), https://maharashtra.gov.in/Site/upload/WhatsNew/SDMP-2015%20(1).pdf

District Nagpur, Government of Maharashtra: https://nagpur.gov.in/demography/, https://nagpur.gov.in/geography-climate/

EcoCity Standards, https://ecocitystandards.org/framework/

European Commission, Smart cities, https://ec.europa.eu/info/eu-regional-and-urban-development/topics/cities-and-urban-development/city-initiatives/smart-cities_en

Google map (Maharashtra, Nagpur) https://www.google.com/maps/place/Maharashtra,+India/@19.5739444,70.5091692,5.11z/data=!4m6!3m5!1s0x3bcfc41e9c9cd6f9:0x1b2f22924be04fb6!8m2!3d19.7514798!4d75.7138884!16zL20vMDU1dnI?entry=ttu

ICLEI-Local Governments for Sustainability, https://iclei.org

ISO 37120; Sustainable Development of Communities—Indicators for City Services and Quality of Life. 1st ed. International Organization for Standardization: Geneva, Switzerland (2014)

E. Wataya and R. Shaw, *Soft Assets in Smart and Resilient City Development*, Disaster Risk Reduction, https://doi.org/10.1007/978-981-97-3373-6

Jiang B (2019) Living structure down to earth and up to Heaven: Christopher alexander, Urban Sci 3:96

Map creation (website) Google map (Odisha, Bhubaneswar) https://www.google.com/maps/place/Bhubaneswar,+Odisha,+India/@19.4950984,71.4062159,5z/data=!4m15!1m8!3m7!1s0x3a226aece9af3bfd:0x133625caa9cea81f!2sOdisha,+India!3b1!8m2!3d20.2375561!4d84.2700179!16zL20vMDFsNjln!3m5!1s0x3a1909d2d5170aa5:0xfc580e2b68b33fa8!8m2!3d20.2960587!4d85.8245398!16zL20vMDNjenFz?entry=ttu

Ministry of Housing and Urban Affairs Government of India, Climate Smart Cities Assessment Framework 2.0. https://smartcities.gov.in/climatesmart_cities

Ministry of Housing and Urban Affairs, Government of India, Smart City Mission (2018) India Smart City Mission. The Smart City Challenge Stage 2—Nagpur, 2018. Available online: https://smartnet.niua.org/sites/default/files/resources/Nagpur_SCP.pdf

Ministry of Housing and Urban Affairs, Government of India, Smart City Mission, ABOUT SMART CITIES MISSION, https://smartcities.gov.in/about-the-mission

Ministry of Housing and Urban Affairs, Government of India, Smart City Mission, SPVs, https://smartcities.gov.in/implementation

Ministry of Housing and Urban Affairs, Government of India, Smart City Mission, FINANCING, https://smartcities.gov.in/financing

Ministry of Housing and Urban Affairs, Government of India, Smart City Mission, Convergence, http://164.100.161.224/content/innerpage/convergence-sp.php

Ministry of Housing and Urban Affairs, Government of India, India Smart City Mission, The City Challenge Stage 2 Bhubaneswar. 2016. Available online: https://smartnet.niua.org/sites/default/files/resources/BHUBANESWAR_SCP_0.pdf

Ministry of Housing and Urban Affairs, Government of India, Smart City Mission, VISION Available online: http://smartcities.gov.in

Ministry of Housing and Urban Affairs, Government of India, Smart City Mission, http://164.100.161.224/content/

Ministry of Housing and Urban Affairs, Government of India, Smart City Mission, IMPLEMENTATION, https://smartcities.gov.in/implementation

Ministry of Housing and Urban Affairs, Government of India, Smart City Mission, Monitoring https://smartcities.gov.in/implementation

Ministry of Housing and Urban Affairs, Government of India, Smart City Mission, Strategy, http://164.100.161.224/content/innerpage/strategy.php

Ministry of Housing and Urban Affairs, Government of India, Smart City Mission, STRATEGY, https://smartcities.gov.in/strategy

Ministry of Housing and Urban Affairs, Government of India (2021) Ease of Living Index 2020: Assessment Report, https://livabilitystore175634-prod.s3.amazonaws.com/public/docs/Ease_of_Living_Report.pdf

Nagpur Municipal Corporation, https://www.nmcnagpur.gov.in, https://www.nmcnagpur.gov.in/demographic-profile

Nagpur Municipal Corporation, https://www.nmcnagpur.gov.in, https://www.nmcnagpur.gov.in/location-and-connectivity

Nagpur government, District Nagpur, https://nagpur.gov.in/history/; https://nagpur.gov.in/geography-climate/

Nagpur Metropolitan Regional Development Authority (NMRDA), http://www.nmrda.org

Nagpur Municipal Corporation Citizen Engagement (2015) Government of India https://blog.mygov.in/nagpur-municipal-corporation-citizen-engagement/

Network Analyzer Online Help (Cytoscape) (2018) Max Planck Institutes, Available online: https://med.bioinf.mpi-inf.mpg.de/netanalyzer/help/2.7/

Nature India, https://www.nature.com/articles/nindia.2021.18

Neirotti P, De Marco A, Cagliano AC, Mangano G, Scorrano F (2014) Current trends in Smart City initiatives: Some stylised facts, Department of Management and Production Engineering,

Politecnico di Torino, Corso Duca degli Abruzzi 24, 10129 Torino (TO), Italy. Cities Int J Urban Policy Planning 38:25–36
OCED (2022) OECD policy paper, 'smart cities and inclusive growth, building on the outcomes of the 1st OECD roundtable on smart cities and inclusive growth', https://www.oecd.org/cfe/cities/oecd-roundtable-on-smart-cities-and-inclusive-growth.htm
Odisha State Disaster Management Authority, https://www.osdma.org/disaster-risk-reduction/#gsc.tab=0
OECD, Resilient cities, https://www.oecd.org/cfe/resilient-cities.htm
Prime Minister's Office of Japan, Japan's Smart Cities - Solving Global Issues Such as the SDGs, etc. through Japan's Society 5.0, Government of Japan, Kantei. https://www.kantei.go.jp/jp/singi/keikyou/pdf/Japan%27s_Smart_Cities-1(Main_Report).pdf
Resilience Alliance, 'Resilience', https://www.resalliance.org/resilience
Sveiby KE (2001) Methods for measuring intangible assets. Updated 2010 http://www.sveiby.com/files/pdf/intangiblemethods.pdf
The Rockefeller Foundation, 100 Resilient Cities, https://www.rockefellerfoundation.org/100-resilient-cities/
The New Indian Express, BMC readies action plan to keep cyclone Amphan at bay https://www.newindianexpress.com/cities/bhubaneswar/2020/may/18/bmc-readies-action-plan-to-keep-cyclone-amphan-at-bay-2144801.html
The National Institution of Urban Affairs, EASE of LIVING 2020, Citizen Perception Survey, https://livabilitystore175634-prod.s3.amazonaws.com/public/docs/Citizen_Perception_Survey_Report.pdf
The Times of India, Homes flooded, traffic hit, few hours of rain inundates Nagpur https://timesofindia.indiatimes.com/city/nagpur/homes-flooded-traffic-hit-few-hours-of-rain-inundates-nagpur/articleshow/84251270.cms
The United Nations Department of Economic and Social Affaires Population Division, https://population.un.org/wup/publications/Files/WUP2018-PopFacts_2018-1.pdf
The United Nations Office for Disaster Risk Reduction (2017) Climate Disconnect in India's Smart City Mission, PreventionWeb, https://www.preventionweb.net/news/climate-disconnect-indias-smart-cities-mission
The United Nations, SDG Indicators, https://unstats.un.org/sdgs/metadata/?Text=&Goal=11&Target=
The United Nations University Institute for Environment and Human Security (UNU-EHS), Bundnis Entwicklung Hilft and the University of Stuttgart in Germany (2020) World Risk Index, https://byjus.com/current-affairs/world-risk-index/
The United Nations, Sustainable Development Goals, Goal 11: Make cities inclusive, safe, resilient and sustainable, https://www.un.org/sustainabledevelopment/cities/
The World Bank, Understanding Poverty Digital Development, Brief, Smart Cities, January 8, 2015, https://www.worldbank.org/en/topic/digitaldevelopment/brief/smart-cities
The World Bank, Understanding Poverty, Urban Development, https://www.worldbank.org/en/topic/urbandevelopment
The World Bank, Data, Urban population, https://data.worldbank.org/indicator/SP.URB.TOTL
The World Bank, Data, Urban population (% of total population) https://data.worldbank.org/indicator/SP.URB.TOTL.IN.ZS
World Economic Forum The Centre for the Fourth Industrial Revolution Japan https://note.com/c4irj/n/nbd1c96c9e811
World Population Review, Bhubaneswar Population 2023 (worldpopulationreview.com) Nagpur Population 2023 (worldpopulationreview.com)

SPRINGER NATURE

GPSR Compliance

The European Union's (EU) General Product Safety Regulation (GPSR) is a set of rules that requires consumer products to be safe and our obligations to ensure this.

If you have any concerns about our products, you can contact us on ProductSafety@springernature.com

In case Publisher is established outside the EU, the EU authorized representative is:

Springer Nature Customer Service Center GmbH
Europaplatz 3
69115 Heidelberg, Germany

The manufacturer's authorised representative in the EU is Springer Nature Customer Service Centre GmbH, Europaplatz 3, 69115 Heidelberg, Germany. If you have any concerns regarding our products, please contact ProductSafety@springernature.com

Printed and bound by CPI Group (UK) Ltd, Croydon, CR0 4YY
01/10/2026
02225852-0011